# A MEDIEVAL FARMING GLOSSARY OF LATIN AND ENGLISH WORDS

taken mainly from
Essex records

COMPILED BY
JOHN L. FISHER

SECOND EDITION REVISED BY
AVRIL & RAYMOND POWELL

*Published by the*
ESSEX RECORD OFFICE
County Hall, Chelmsford, Essex CM1 1LX

© Essex County Council, 1997

All rights reserved.
This book may not be reproduced, in whole or in part,
without written permission from the publishers.

*British Library Cataloguing in Publication Data*
A catalogue record for this book is available from the British Library

ISBN 1 898529 13 2

Essex Record Office Publication No. 138

*Cover illustration:*
*The image used on the cover is re-drawn from MS Bodl. 764 fol. 41v,*
*reproduced by courtesy of the Bodleian Library, Oxford*

Printed by ESSEX *Print* & GRAPHICS (01245) 450777, part of Essex County Council, from digital media supplied

# EDITORIAL NOTE

John Lionel Fisher F.S.A., priest, sportsman and scholar, was born in York. He won an exhibition to Sidney Sussex College, Cambridge, graduating in 1908. Ordained in 1911, he held curacies in Hertfordshire (1911-18). While teaching during the First World War he met L.F. Salzman (later General Editor of the *Victoria County Histories*), who kindled his interest in local history. After the War he became rector of Netteswell in Essex (1918-54, with Little Parndon, 1921-54), and later rural dean of Harlow and honorary canon of Chelmsford. He was a leading member of the Essex Archaeological Society, contributed many papers to its *Transactions,* and served for some years as the society's honorary archivist. His independent publications included *The Deanery of Harlow* (1922) and *Harlow New Town: a short history of the area* (1951). He also wrote several articles for the *Victoria County History of Essex,* volumes five and eight. John Fisher also excelled on the sports field, playing hockey and rugby football into his forties. He was, says the *History of the Diocese of Chelmsford,* 'full of information about every town and village in the county [and] equally at home and well-informed at Twickenham, Lords and Wimbledon.'

Canon Fisher started compiling this *Glossary* about 1945 with no thought of publication. An early draft was placed in the Essex Record Office, where it was often consulted by staff and students. At Canon Fisher's request we undertook to arrange for its publication and to see it through the press. It was published in 1968, for the Standing Conference for Local History, by the National Council of Social Service. Canon Fisher, who had himself contributed to the costs of printing, died early in 1969.

On the first edition of the *Glossary* Canon Fisher commented 'I can't see it having much of a sale - the market is so limited.' He was too modest, for the book went out of print within a few years. It proved particularly useful to students of local history and their teachers, including Mrs Jo Ann Buck and Mr David Dymond who urged us to prepare a new edition, and to whom we are grateful for their encouragement. The revised edition is published by the Essex Record Office with the consent of the British Association for Local History, successors to the Standing Conference for Local History.

In the first edition of the *Glossary* about two-thirds of the entries were Latin and one-third Middle English. In preparing the Latin entries for the second edition we have omitted some words which were incorrectly defined in the first edition, or had been taken from post-medieval sources, or would rarely be found in a farming context. But we have added a larger number because they were often used in agriculture *(e.g.* **ager**) or are easily confused with words of the same spelling *(cf.* **acer 1, 2, 3**). Some revision was also necessary in defining the English words, but since they were already generously represented in the *Glossary* in relation to their frequency of occurrence in medieval manorial documents, we have not added substantially to their number.

In the first edition of the *Glossary* the sources of many of the words, though not all, were indicated by means of a code. The references to printed sources were traceable, but those to manuscript sources were so imprecise as to be of doubtful value in locating particular words in their contexts. In the revised edition, therefore, we have omitted the coded references, while extending the other sections of the Bibliography.

The first edition of the *Glossary* was examined in typescript by the late Mr R.E. Latham, editor of the *Medieval Latin Dictionary*, who made many valuable suggestions. His notes have frequently been consulted in compiling the second edition. The present editor of the *M.L.D.*, Dr D.R. Howlett, allowed us to consult unpublished material in the files of the *Dictionary*, and the assistant editor, Dr J. Blundell, answered a number of queries. Mr Peter Northeast and Mr Dennis Knox sent us some words from Suffolk records. Mrs Helen Coghill, with assistance from Mrs Pamela Studd, put the complicated text of the *Glossary* onto computer disk. The County Archivist, Mr Ken Hall, besides accepting the *Glossary* for publication in the Essex Record Office series, has given much support and encouragement to the work. Miss Janet Smith, Assistant County Archivist, has been responsible for seeing the book through the press. To all these friends we express our warmest thanks.

The *Glossary* is based on original sources, mostly manorial records relating to Essex. While it makes no attempt to be comprehensive even within the limits set by its title, it is the fruit of many years' research by a scholar whom we knew as a friend, and in revising it we have tried to preserve its original character.

# ARRANGEMENT OF ENTRIES

The Latin words to be explained are printed in **bold face**, the Middle English words in ROMAN CAPITALS. The entries consist of key-words, followed in many cases by variant spellings. Entries are in strict alphabetical order. When a large number of words with the same root occurs they have usually been separated for ease of reference (*e.g.* **carretta, carrettalis** *et sqq.*) When a word occurs in both Latin and Middle English forms the Latin form is taken as the key-word, even when the Latin word is derived from Middle English (*e.g.* **acremannus**). Verbs are entered in the present indicative, followed in some cases by participles (*e.g.* **caedo**). Adjectives of different gender are listed in the same entry (*e.g.* **certus**).

It must be emphasised that many variant spellings, other than those listed here, may be found in original documents, and in particular that the *Glossary* does not include the variant endings **-cio** and **-cium** for such words as **adunatio** and **capitium**. A variant spelling very different from the key-word of the entry in which it occurs is cross-referred from its own alphabetical place (*e.g.* **cindula**). There are cross-references also from adjectives (or qualifying nouns) to associated nouns (*e.g.* **durus; bidens**).

When an entire word is repeated later in the entry it is usually represented by a single letter (*e.g.* **auca**). An incomplete key-word followed by a dash denotes a general cross-reference to variant spellings (*e.g.* **care**-; **dagg**-;). Words of identical form but different origin have been put in separate entries with serial numbers (*e.g.* **acer 1; acer 2; acer 3**).

# BIBLIOGRAPHY

The words in this *Glossary* are taken from records, mostly manorial, compiled between 1100 and 1600. Most of the words earlier than 1300 were taken from the sources in **List (a)** below, while most of those after 1300 come from sources in **List (b)**. Some of the records, in both lists, were transcribed by Canon Fisher for the first edition of the *Glossary*, and his transcripts are now in the Essex Record Office (reference T/P 145). **List (c)** comprises general works used to check or clarify entries in the *Glossary*.

### List (a) Various National and Local Sources

British Library, Cott. MS. Tib. C. ix (Waltham Abbey rental c. 1235).
British Library, Add. Ro. 28792 (Estreat Rolls 1382).
British Library, Add. Ro. 56455 (Wimbish court rolls, cf. E.A.T. **n.s.** xxi. 330).
E.R.O., Colchester Acc. C38 (Leger Book of St. John's Abbey, Colchester: cf. *E.A.T.* **n.s.** xxiv. 77).
Public Record Office, SC2/173/30 (Waltham Abbey court roll 1270-71).
P.R.O., SC6/849/11 (Dagenham survey 1321)
Westminster Abbey Muniments.
*Cartularium Monasterii Sancti Johannis Baptiste de Colecestria*, ed. S.A. Moore (Roxburghe Club, 1897).
*Catalogue of Ancient Deeds in the Public Record Office*, vol. vi.
*The Domesday of St. Paul's*, ed. W.H. Hale (Camden Soc. 1858).
*Feet of Fines for Essex* (Essex Arch. Soc.), vols. i-iii (1899-1949).
*The Oath Book or Red Parchment Book of Colchester*, ed. W.G. Benham (1907).
*The Transactions of the Barking and District Archaeological Society* (1935-7).
*The Transactions of the Essex Archaeological Society* (1858-1960).

### List (b) Manorial Records in the Essex Record Office

The Essex Record Office has a large collection of manorial records, which can be accessed through the Principal Parish Index. Those most extensively used for the *Glossary* relate to the following places, to which we have added, in brackets, the catalogue marks of the main series to which the items belong.

Birdbrook (D/DU); Blackmore (D/DK and D/DHt); Canfield, Great (D/DMw); Clavering (D/DP); Colne, Earls (D/DPr); Copford (D/DHt); Dunmow, Great (D/DB and D/DMg); Felsted (D/DSp); Foulness (D/DHt); Fyfield (D/DCw); Harlow (D/DEs); Hatfield Broad Oak (D/DBa; D/DC; D/DGe; D/DK; and D/DQ); Havering (D/DSg; D/DU); Hutton (D/DH); Kelvedon (D/DU); Langenhoe (D/DC and D/DEl); Messing (D/DH); Moulsham, in Chelmsford (D/DM); Purleigh (D/DVo); Roding, Abbess (D/DHf); Stock (D/DP); Thaxted (D/DHu); Tollesbury (D/DK); Totham, Great (D/DC); Waltham, Great (D/DTu); Wethersfield (D/DFy); Writtle (D/DP).

**List (c) General Sources used in editing**

*Revised Medieval Latin Word List,* ed. R.E. Latham, 1965.
*Medieval Latin Dictionary,* ed. R.E. Latham and D.R. Howlett, 1975- , in progress.
*Middle English Dictionary,* ed. H. Kurath, S.M. Kuhn, and J. Reidy. 1952- , in progress.
*Oxford English Dictionary.* Microprint edn. 1971.
W. Cunningham, *Growth of Industry and Commerce.* 3rd edn. 1896. (Appendix).
N. Neilson, 'Customary Rents', in *Oxford Studies in Social and Legal History,* ed. P. Vinogradoff. 1910.
J.R. Hulbert, 'English in Manorial Documents of the Thirteenth and Fourteenth Centuries', *Modern Philology,* Aug. 1936.
*Charters and Custumals of the Abbey of Caen,* ed. M. Chibnall. 1982.

# ABBREVIATIONS

| | |
|---|---|
| *adj.* | adjective |
| *cf.* | compare |
| *coll.* | collectively |
| *E.A.T.* | *Transactions of the Essex Archaeological Society* |
| *e.g.* | for example |
| esp. | especially |
| *f.* | female |
| *i.e* | that is |
| *pl.* | plural |
| *p.p.* | past participle |
| *ref.* | reference |
| *sb.* | substantive |
| us. or *us.* | usually |
| *vb.* | verb |
| w. | with |
| † | doubtful meaning |

# MEDIEVAL FARMING GLOSSARY

## A

**abellus**, ABELL, white poplar
**ablactatio**, weaning
**ablacto**, to wean
**abradico**, to uproot
**abrado**, to scrape
**abrasio**, shaving; chip
**abscido**, to cut off; to lop
**abscisio**, cutting off; lopping
**abuttamentum**, abutment
**abutto**, to adjoin; to abut
**accrochiamentum**, encroachment
**accrochio, accrocho**, to encroach upon
**accumulo**, to heap up, to stook (corn)
**acer 1**, sharp; sour; keen
**acer 2**, maple
**acer 3, acerum, asser**, steel
**acermannus** *see* **acremannus**
**acero, ascero**, to tip w. steel
ACOURWARE *see* **acrewara**
**acra**, acre; strip of ploughland
ACRELONDE, ACRELONGE, smallholding, acreland *cf.* AKERMANLONDE
**acremannus, acermannus**, AKERMAN, AKYRMAN, smallholder, acreman
**acrewara**, ACOURWARE, ACREWARE, ACURWARE, (?) acre liable for geld; *cf.* **wardacra**
**acuatio, acuitio**, sharpening, whetting
**acuo**, to sharpen; to whet
ACURWARE *see* **acrewara**
**acus 1**, husk, chaff
**acus 2**, pin; needle
**adjunctio, adjunctura**, drafting (esp. of animals into next age-group)
**adjungo**, to yoke; to attach; to draft (animals)
**adunatio**, gathering up
**aduno**, to rake; to gather up
**aeneus, eneus**, of bronze
**aequo, equo**, to make level or equal; to balance (account)

**aes, es,** copper; brass; bronze
**aestas, estas,** summer
**aestivalis, estivalis,** of or for summer
**aestivatio, estivatio,** summer pasture
**aetas, etas,** age (of person or animal); *see also* **porcus**
**affortio** *see* **effortio**
**affra, affrus** *see* **averus**
**affraia** *see* **effraia**
**ager**, field
**agger**, heap; stack; embankment
**agistamentum, ingistiamentum,** agistment, (right of or payment for) pasture
**agisto**, to assess (animals or woodland) for pasture due; to put to pasture
**agnellatio**, lambing
**agnello**, to lamb
**agnellus**, lambkin
**agnus**, lamb, **agna**, ewe lamb; **agnus datus S.Antonio**, 'tantony', runt offered to St. Antony
**agricola**, farmer; tiller of soil
**agulettus**, metal tag
**aisiamentum, aysiamentum, heisiamentum**, easement
AKERMAN *see* **acremannus**
AKERMANLONDE, smallholding; *cf.* ACRELONDE
AKYRMAN *see* **acremannus**
**albus**, white; **album**, dairy produce; *see also* COORIER; **firma**
ALEBEDREP, boonwork w. ale
ALEE, alley
ALEFOUNDER, ALEFUNDER, ale-taster
ALETOL, toll on ale
**algea, alga, algeus**, trough; vat
**allec**, herring
**alleum** *see* **allium**
**alligo**, to bind; to fasten
**allium, alleum**, garlic
**allocamentum**, allowance

1

# ALLOCO

**alloco**, to allow (in account); to hire
**almarium** *see* **armarium**
ALMESSEHOUSE, almshouse
**alnetum**, alder-holt
**alnus**, alder
**altilia** (*pl.*), (fattened) poultry
**altus**, high
**alveus, alvea, alveum**, trough, tub
AMAD, (?) hay meadow
**ambitus**, precinct; region
**amerciamentum**, amercement, fine
**amercio**, to assess for fine
**amor**, love; *see also* **precaria**
**amputatio**, felling
**amputo**, to fell; to cut off; to mow
**anas**, duck
**ancella, ancera**, AUNCER, balance
**ancer** *see* **anser**
**ancera** *see* **ancella**
**ancilla**, maidservant
**androchia**, dairymaid
ANEVELSTOK, anvil block; *see also* ANVELLE
**Anglia** *see* **urbanitas**
**anguilla**, eel; *see also* **avalatio**
**angulus**, corner
**animal**, beast, head of cattle; **animalia juncta**, plough beasts
**annona**, grain; crop
**annualis**, annual, yearly; (of animal) one year old
**annulatio** *see* **anulatio**
**annus**, year
**annuus**, yearly
**ansa**, handle
**anser, ancer**, goose; gander
**antica**, front door or gate
**Antonius** (Saint), *see* **agnus**
**anulatio, annulatio**, ringing (esp. of pigs)
**anulo**, to fit w. a ring
**anulus**, ring
ANVELLE, anvil; *see also* ANEVELSTOK
**aper**, boar
**aperculus**, young boar
**aperio**, to open, to unblock; **apertus**, open; *see also* **tempus**
**aperiolus** *see* **apriolus**
**apertio**, opening; clearing

**apis**, bee; *see also* **examen; vasum**
**apium**, parsley
**apparatus**, fittings, esp. harness
**appediamentum** 'foot', support (of dam)
**approuamentum, appruamentum**, source (of profit)
**approuo**, to make a profit (esp. by reclaiming land)
**apriolus, aperiolus**, young boar
**aptatio**, fitting; dressing (of timber)
**apto**, to fit; to fashion
**aqua**, water
**aqualitium**, gutter
**aquarium**, watering-place; meadow
**aquaticus** *see* **lupus; sulcus**
**aquo**, to water (animal)
**arabilis**, arable
**aratralis**, concerned with ploughing
**aratrum**, plough; ploughland
**aratura, arrura**, ploughing (service)
**arca, archa**, coffer; chest; fish-trap
ARCHERESELVER, (?) payment in lieu of providing archer
**architector, architectus**, builder; thatcher
**arcus**, bow; bucket-handle
**areragium** *see* **arreragium**
**aresta** *see* **arista**
†**argento** *see* **arrento**
**argentum**, silver; **a. vivum**, mercury
**argilla**, clay
**aries**, ram
**aripennis**, ARPENT, measure of land, esp. of vineyard, arpent
**arista, aresta**, 'beard', ear (of corn)
**armariolum**, small cupboard
**armarium, almarium**, cupboard
**armentum**, (herd of) cattle
**arnementum, arnamentum**, salve, 'vitriol'; *cf.* **atramentum**
**aro**, to plough; *see also* GAVEL
ARPENT *see* **aripennis**
**arrento**, †**argento**, to rent out; to commute (service)
**arreragium, areragium** (*us. pl.*), arrears
**arrura** *see* **aratura**
**artavus**, knife
**arto**, to narrow; to constrict

2

**ascaldaria** *see* **escaldaria**
**ascero** *see* **acero**
**asperduta** *see* **esperduta**
**asporto**, to take away; to steal
**assarta** *see* **essarta**
**asser 1** *see* **acer 3**
**asser 2**, board, plank
**assisus** *see* **redditus**
**asso**, to roast
**astella, hastella**, stick; shaft
**astringo**, to confine; to make narrow; **astrictus**, narrowed
**astringo**, to confine; to make narrow; **astrictus**, narrowed
**astrum 1**, star
**astrum 2**, hearth; dwelling
**asturcus, osturcus**, sparrow-hawk
**atramentum**, salve, 'vitriol'; *cf.* **arnementum**
**attachio**, to attach; to fasten; to tie; to bind (person) to appear in court
**attastator**, taster (esp. of ale)
**auca**, goose; **a. mariola, marola**, (?) unmated goose
**aucella, aucellus, aucula**, gosling
**auctionator, auctionatrix, auxionator**, retailer
**aucupatio**, wildfowling
**aufugo**, to drive away
**aufugio**, to run away
**augmentatio, augmentum, aumentatio**, enlargement; addition
**Augustus** *see* **gula**
**aula**, hall; manor house
**aumentatio** *see* **augmentatio**
AUNCER *see* **ancella**
AUNDERNE, AWNDERNE, andiron
**auragium** *see* **averagium**

**auricula, auriculus**, plough-ear; lug
**autumnalis, autumnus** *see* **custos**
**auxionator** *see* **auctionator**
**avalatio (anguillarum)**, (catching of) eels during migration downstream
**avantagium**, profit; difference between razed and heaped measure
**avena**, oats
**avenarius, avenator**, 'avener', purveyor of oats
**averagium, auragium**, carrying service or payment in lieu; **a. pedestre**, carrying service on foot
**averium**, property; livestock; head of cattle
**avero**, to perform carrying service
AVERSELVER, payment in lieu of carrying service
**averus, affra, affrus, avera, haverus**, draught beast
**avesabilis**, liable for pannage
**avesagium, avisagium**, (payment for) pannage; *cf.* GRASAVESE
**aveso**, to pay pannage
**avis**, bird
AWNDERNE *see* AUNDERNE
**axa 1**, axe
**axa 2**, axle
**axatio**, fitting w. axle
**axelo, axo**, to fit w. axle
**axis**, axle (of wheel); spindle (of millstone)
AXNAIL, EXNAIL, axnail, nail for fastening axle-tree to cart
AXTRE, axle-tree
**aya** *see* **haia 1**
AYBOTE *see* HAYEBOTE
**aysiamentum** *see* **aisiamentum**

# B

BAAZ *see* **bassum**
**bacinus**, basin
**baco**, bacon; *see also* **perna**
**bacropa**, BAKROPE, back-rope of cart
**baculus**, cudgel; staff; part of structure
**badius, basius, bayus**, bay (horse)

**baga, bagga**, bag, wallet
**baia 1, baya**, bay of pond, mill-dam; *see also* FORBAY; HENDERBAY; REREBAY
**baia 2, baya, bayus**, bay, compartment of building
**baillivus, ballivus**, bailiff; agent

**baissa, bayssa,** maidservant
**baitatio, bartatio,** 'baiting', provision (for horse)
BAKBERANDE, (thief) caught w. stolen property
BAKHOUS, bakery
BAKROPE *see* **bacropa**
BAKYREN, wrought iron or (?) supporting irons at back of cart
**balcus,** ridge separating fields; balk, side-beam
**ballista,** crossbow
**ballivus** *see* **baillivus**
**bancarium,** BANKER, seat cover
**banchus** *see* **bancus 2**
**bancus 1,** bank, mound
**bancus 2, banchus, bancum,** bench; seat; **b. tornatile** revolving seat; **liber b.,** free bench, widow's right of dower
BANKER *see* **bancarium**
**banleuca, benleuca,** banlieu, area of jurisdiction
**baratator, barector,** trouble-maker
**barbatio,** setting (of fence) w. thorns
**barbo,** to set w. thorns; *cf.* **berdo**
**barcaria** *see* **bercaria**
**barcarius** *see* **bercarius**
**barector** *see* **baratator**
**barellus, barrillus,** barrel, cask; part of structure
**bargia,** saddle-pad; collar-pad
**barhudum,** BARHEDE, BARHIDE, cart-cover
BARKER, tanner
BARLINES (*pl.*), traces, part of harness
BARNTON *see* **bertona**
**barra,** bar, barrier
**barrillus** *see* **barellus**
**bartatio** *see* **baitatio**
BARTON *see* **bertona**
**basius** *see* **badius**
**bassum, †basta,** BAAZ, saddle-pad; collar-pad
**bassus,** low; *see also* **camera**
**†basta** *see* **bassum; bastum**
**bastum, †basta,** 'bast', bark fibre (rope)
**batellagium,** boat hire or (?) service
**batellus, batella, batellum,** small boat
**bateria, baterissa,** threshing-barn
**bavenum,** BAVEYN, 'bavin', bundle of sticks
**baya** *see* **baia 1; baia 2**
**bayssa** *see* **baissa**
**bayus 1** *see* **badius**
**bayus 2** *see* **baia 2**
**†bedacra,** service of (?) reaping or mowing one acre; *cf.* BEDHALFAKER
BEDEHALFAKER *see* BEDHALFAKER
**bedellaria, bedelleria,** office of beadle
**bedellus,** beadle
BEDEMAD, mowing service
**bederipa,** BEDERIP, BEDRIP, reaping service; *cf.* ALEBEDREP; CUSTUMBEDREPPE; SOKNEBEDRIPE; WATERBEDRIP
BEDERIPDAYES, reaping service
BEDERIPSELVER, payment in lieu of reaping service
BEDERODE, service of (?) reaping and binding a rood of corn
BEDESEL, (?) ploughing service
BEDEWEDE, BEDEWEDYING, hoeing service
BEDHALFAKER, BEDEHALFAKER, service of (?) mowing half an acre; *cf.*†**bedacra**
BEDHALFAKYRMAN, tenant holding by the above service
BEDRIP *see* **bederipa**
BEEMFELLING, beam filling, plasterwork between rafters and roof
BEER, bier
BELETTE *see* **billettum**
BELLEFLES, belly-fleece; *see also* **lana**; WAMBELOKES
BELLES, belly-hide
BELLEWETHYR, bell-wether, leader of flock
**bena,** BENE, boon-work
**benda,** (metal) band
BENE *see* **bena**
**benedico,** to bless; *see also* **panis**
BENEMONG, mixed crop, including beans; *cf.* PESEMONG
**benertha,** BENERTHE, boon-ploughing
**benleuca** *see* **banleuca**
BENSCHEF, sheaf, perquisite of reapers; *cf.* MENESCHEF

**bera 1**, BERE, beer
**bera 2**, BERE, bear barley; *cf.* **hastiberum**
**bercaria, barcaria**, sheep walk or pen
**bercarius, barcarius, bercharius**, shepherd
**berdo**, (?) **berto**, to set a fence w. thorns; *cf.* **barbo**
BERE *see* **bera 1; bera 2**
**berewica**, 'berewick', outlying part of manor
BERNEHAWE, barnyard
**berso**, to shoot (game)
**berto** *see* **berdo**
**bertona**, BARNTON, BARTON, manorial enclosure
BERTYNGNAYL, (?) floor-nail
**besca**, spade
**bestia**, (farm) animal
BETER, BETERE, instrument for beating
BETILL, BETEL, mallet
BEYL *see* **builum**
**biberium**, drinking time; (?) drink money
**bibo**, to drink
**bidens**, sheep; *see also* **campana**
**biga**, cart; **b. ferrata**, iron-tired cart
**bigalis**, for a cart
**bigarius**, carter
**bigata**, cart load
**bilegium** *see* **birelagia**
**bilettus** *see* **billettum 1**
**billa 1**, bill, billhook; steel tool for dressing millstone
**billa 2**, label; receipt; informal document; **b. indentata**, indenture
**billettum 1, bilettum**, BELETTE piece of (fire) wood
**billettum 2**, note, receipt
**birchettum**, birch-grove
**birelagia, bilegium**, by-law, local regulation
**birretum, birettum**, cap
**bissus**, brown (bread); bay (horse)
**bladum**, corn, grain; **b. durum**, hard corn; **b. molle**, soft corn
BLANCHEFERME, blanch farm, rent paid in silver; *see also* **firma**
**blanchettus, blankettus**, white woollen cloth

**blodius**, blue
**bluettus**, blue (cloth or colour)
BOBBYNGRODE, (?) service of binding one rood of corn
BOCAT, BOCATTE *see* **bukettus**
BOCE, boss, knob
BODITRAY, BODYTRAY, handle or sideboard for body of cart
BOKET *see* **bukettus**
BOLEMONG *see* BULIMONG
**boletellus** *see* **buletellus**
**bolla**, bowl; liquid or dry measure
**bollo**, (?) to poll, to reduce (tree) to a bole
BOLSTER, BOLSTRE, (metal) support or cushion
**bondagium**, bondage, villeinage; tenure or tenement so held
BORCHALPENING *see* BORGHALPENI
**borda 1**, board, plank
**borda 2**, cottage
**bordarius, bordar,** smallholder
BORDCLOTH, table-cloth
**bordellum, bordellus**, smallholding
BORDELOND, smallholding
**bordmannus**, smallholder
**bordnailum**, BORDNAYL, board nail
**bordo, burdo**, to fit w. boards
BORE *see* BOUR
BORGHALPENI, BORCHALPENING, (?) fine paid at view of frankpledge; *cf.* BOURGHSELVER
BORIL, (?) hard wood (for arrow)
**borla** *see* **burla**
**bortreminga**, view of frankpledge
**bos**, ox
**boscus**, wood; woodland; **b. forinsecus**, outlying wood; *see also* UTWODE
BOSSYNG *see* BUSSHING
**bota**, BOTE, boot
**botellus**, bundle (of hay)
BOTELMAKER, BOTTLEMAKER, hay presser
BOTORE *see* **butor**
BOUENET, BOWNETTE, fishtrap of wickerwork
BOUGE, opening (in roof of dovehouse)
BOUK, BUC, BUCK, body (of wagon)
BOULLS *see* CLOSHE BOULLS
BOUNDALL *see* **bunda**

BOUR, BORE, cottage; shelter; bower
BOURGHSELVER, (?) tithing payment; *cf.*
    BORGHALPENI
**bovaria, boveria**, cow-shed
**bovarius**, oxherd; cowherd
**bovettus**, steer, bullock
**bovicarius, bovicularius**, bullock-keeper
**boviculus**, bullock
BOWLING, pollard tree
BOWNETTE *see* BOUENET
**boxo**, to fit (wheel) w. axle-box
BRAAZ, BRAS, (?) brace, arm (of mill-axle)
**bracae, braccae**, breeches
**bracchium, brachium**, (?) **braga**, arm; bar; band
**bracha** *see* **brecca 2**
**brachium** *see* **bracchium**
**braciator, braciatrix, brasiator**, brewer; ale-wife
**bracinum, bracinium**, brewhouse; brewing
**bracio, brasio**, to brew
**bracium, brasium**, malt
**braga** *see* **bracchium**
BRANK, buckwheat
BRAS *see* BRAAZ
**brasiator** *see* **braciator**
**brasio** *see* **bracio**
**brasium** *see* **bracium**
**brecca 1**, BRECHE, BREKKE, breach (esp. in dam)
**brecca 2, bracha, brecha**, BRECHE, assart, clearing
BRECHE *see* **brecca 1**; **brecca 2**
BREDEGRATE, (?) bread-grater
BREKKE *see* **brecca 1**
BRESTCLOUTE, metal patch on breastboard of plough
BRICKKILL, brick kiln
**broca 1**, brook; watermeadow
**broca 2, brocha, brochia**, BROCHE, fastening (for sack); skewer
BROCHE *see* **broca 2**; **brusca**
**broddum**, brad-nail
BRODNAYL, BRODHEDNAYL, brad-nail
BROWSYNGWODE, (?) brushwood

**bruaria, bruarium, brueria**, heath; heathland
**bruillus, brulla**, thicket, coppice
**brusca, brusa**, BROCHE, scrub, brushwood
BRYKMAN, brickmaker
**bubalus, bubulus**, bull; ox
**bubulcus**, oxherd or swineherd
BUC, BUCK *see* BOUK
**buckettus** *see* **bukettus**
**buffetum, bufetum**, stool; sideboard
**builum**, BEYL, BUIL, curved handle of plough
**bukettus, buckettus**, BOCAT, BOCATTE, BOKET, bucket; bucketful
**buletellus, boletellus, bultellus**, bolting-cloth; sieve
BULIMONG, BOLEMONG, BULLYMONDE, BULLYMONG, mixed crop for fodder
**bultellus** *see* **buletellus**
**bunda**, BOUNDALL, boundary (mark)
**bundellus, bundillus**, bundle
BURAT, (?) container
**burdo** *see* **bordo**
**burellus**, 'burel', coarse cloth
**burgagium**, borough tenement or tenure, burgage
**burgus**, borough
**burla, borla**, wool, flock
BURLE, to dress nap of cloth
**burrochius**, BURROK(E) fish-trap
**bursa**, purse, bag; *see also* **portator**
**busca, buscha**, firewood
**bussellus**, large bowl, container; dry measure, bushel
BUSSHING, BOSSYNG, BUSHYNGE, (gathering of) firewood
BUTALLS, BUTTOULES, abutments
**butor, butus**, BOTORE, bittern
**butta 1**, butt, small strip of land
**butta 2, buttus**, cask, butt
**butto**, to abut
BUTTOULES *see* BUTALLS
**butus** *see* **butor**
**butyrum**, butter
BYCHE, bitch
BYNDE, binding-tie

# C

**caballarius** *see* **molendinum**
**caballus**, horse
**cablicium**, windfallen wood
**caccabus, cacabus**, cauldron
**cacia, chacia**, hunting; drove-way; right of way
**cacio, chacio**, to hunt; to drive (cattle)
**cacior, chascurus**, hunter (horse); drover
**cadaver**, carcase
**cadaverator**, inspector of carcases
**cadus**, CADE, cask; barrel
**caecus, cecus, secus**, blind
**caedo, cedo**, to kill; to cut down; **caesus**, hewn (of wood)
**caeduus, ceduus**, coppice, wood grown for periodical cutting; *see also* **silva**
**caenovectorium, cenevectorium**, dung-cart; (wheel) barrow
**caepa, cepa**, onion
**caesus** *see* **caedo**
CAGGE, CAIGE, cage, lock-up
**caium** *see* **kaya**
**calamo, calmo**, to gather stubble; *cf.* **halmo**
**calamus**, reed; stalk; pipe
**calathus**, vessel; basket
**calcar**, spur
**calcea** *see* **calceta**
**calceatura, calcatura, calciatura**, footwear
**calceta 1, calcea, calcetum**, CAUCEE, CAUNSY, KAWSE, causeway
**calceta 2, calcetum**, lime, whitewash
**calco**, to tread (grapes, or hay in stack)
**caldarium, chalderum**, cauldron; measure for meal, chalder
**calefacio**, to heat; to scald; to ferment (grain)
CALFAT, to caulk
CALFTRE, (?) wooden appliance associated with a millstone
**caligae** (*pl.*), hose, stockings
**calmo** *see* **calamo**
CALTRE, halter
**camera**, room, chamber; **c. bassa**, room on ground floor; **c. privata**, privy

**caminus, camina**, CHEMENEYE, hearth, fireplace
**camisia**, shirt
**campana**, bell; **c. bidentium**, sheep-bell
CAMPEN, CAMPING, ball-game (football or form of hockey); C. CROOKE, hooked stick used in this game
**campus**, field, cultivated field
**canabus, canapus** *see* **cannabis**
**candela**, candle; **c. de** COTOUN, candle w. cotton wick
**canellus**, watercourse; gutter
CANEVAS *see* **cannabis**
CANGLE, CANGYLL, (?) enclosure; (?) *cf.* **cantellus 1**
**cannabis, canabus, canapus**, CANEVAS, hemp, canvas
**cantellus 1, cantellum**, 'cantle', small portion; corner; (?) *cf.* CANGLE
**cantellus 2, cantellum**, difference between razed and heaped measure (of grain)
**canthus, cantus**, rim, felloe (of wheel)
**capella, capellum**, hood; cap; chaplet; chapel
**caper**, he-goat
**capister** *see* **capistrum**
**capistro**, to fit w. halter
**capistrum, capister**, halter, headstall
**capitalis**, principal, chief (esp. of officials); *see also* **plegius**
**capiterra, capitera**, headland
**capitium 1, caputium**, hood
**capitium 2**, headland; strip of meadow
**capito**, to fit (plough) w. a head; to abut on
**capo, chapo**, capon
**cappa**, cape; cloak; cap; (iron) cap of harrow etc.
**capra**, she-goat
**caprea, capreola**, doe
**capreus, capreolus**, roebuck
**caput**, head; top; tip (of land etc.); *see also* **scotum**

**caputium** *see* **capitium 1**
**carata** *see* **carrata**
**carbo**, charcoal; **c. maritimus**, sea-coal
**carcatio, cartatio**, loading
**carco**, to load, to stow
**carculo** *see* **sarculo**
**cardo 1**, hinge; piece of iron
**cardo 2**, thistle, teasel
**care-** *see* **carre-, carri-**
**carect-, caret-** *see* **carrett-**
**caritas**, charity; dearth; high price
**caritative, karatative**, charitably
CARLOKKE, KARLOKE, charlock
**carnettus**, GARNET, GERNETT, hinge
**carrata, carata**, cart-load
**carrea, carreum**, cart-load
**carreium**, carriage, (right of) transport
**carretta, carecta**, cart; **c. curta, courta**, short cart; **c. ferrata**, iron-tired cart; **c. nuda**, wooden-tired cart; *see also* **equus; scala; via**
**carrettalis, carectalis**, for or belonging to a cart or carter
**carrettarius, carectarius**, for or belonging to a cart or carter
**carrettata, carectata, caretata, karectata**, cart-load
**carriagium**, transport; carriage service
**carriatio**, carting
**carrio, careo, carrico, charchio, charcio, kario**, to cart, to carry
**carriotta, chariettum**, wagon, carriage
**carruca, caruca**, plough; plough-team; *see also* **fugator**
**carrucagium**, tax on plough-team
**carrucarius, karukarius**, for or concerned with ploughing; ploughman
**carrucata**, carucate, (measure of) ploughland
**carrucator, carucator**, ploughman
**carrus**, cart, wagon; measure of weight; char (esp. of lead)
**cart-** *see also* **carc-**
CARTCLUT, iron plate protecting axle-tree
CARTEBOTE, allowance of wood for repairing carts
CARTGAPPE, cart-way through hedge
CARTLADDER, removable framework on body of cart

CARTLOND, holding, tenement
**caruc-** *see* **carruc-**
**caryophyllum, gariofilum, gariophilum**, clove; *see also* **clavus**
**caseus, casius**, cheese; *see also* **domus**
CASTELWARDE, castle guard service
**castigatorium**, instrument of punishment; (?) whipping post
**castricius, castritus**, castrated animal, wether
**catallum, katallum**, chattel, property, esp. cattle
**catena, cathena, chathena**, chain
**cathedra**, chair, seat
**cathena** *see* **catena**
CATLACCHE, (?) cat-flap (in door)
**cattus, catus**, cat; **c. silvestris**, wild cat
**catulus**, puppy; cub
**catus** *see* **cattus**
CAUCEE *see* **calceta 1**
**cauda**, tail; **caudae porcorum**, pigs' tails, perquisite of swineherd
CAUNSY *see* **calceta 1**
**cavilla**, peg, pin; tooth (of harrow)
CAYE *see* **kaya**
**cebum** *see* **sebum**
**cecus** *see* **caecus**
**cedo** *see* **caedo**
**ceduus** *see* **caeduus**
**celda** *see* **selda**
**cella 1**, cell, compartment
**cella 2** *see* **sella**
**cellarium**, cellar, storeroom
**celtis, seltis**, chisel
**cendula** *see* **scindula**
**cenevectorium** *see* **caenovectorium**
**censarius, censuarius**, rent-paying (tenant)
**census**, tax; rent
**cepa** *see* **caepa**
**ceparatio** *see* **separatio**
**cepes** *see* **sepes**
**cepum** *see* **sebum**
**cera 1**, wax, esp. for candles; beeswax; *see also* **denarius**
**cera 2** *see* **sera**
**ceragium**, wax-shot, payment to church for candles
**cerasum, ciresum**, cherry

**cercella**, teal
**cerculus** *see* **circulus**
CERFTREE, SORFTRE, service tree, rowan
**cerotheca** *see* **chirotheca**
**certus**, fixed, certain; **certum**, fixed payment, service or allowance; **certum letae**, fine at court leet
**cervisia, cervisa**, ale; *see also* **tastator; tentator; tipulator**
**ceuda** *see* **selda**
**chacia** *see* **cacia**
**chacio** *see* **cacio**
**chalderum** *see* **caldarium**
CHALFHOPE, CHALFPETTLE, CHALFPIHTLE, enclosure for calves
**chalo**, blanket
**chapo** *see* **capo**
**char-, charc-, charch-** *see* **carr-**
**chargerium**, CHARGERE, large dish, charger
**chariettum** *see* **carriotta**
**chascurus** *see* **cacior**
CHASHOWS, CHESHOWS, cheesehouse; *see also* **domus**
CHASLEPE, cheeselip, cheesebag
**chathena** *see* **catena**
CHEKER, counting-board; chess-board
CHEMENEYE *see* **caminus**
**cheminagium, chiminagium**, 'way-penny', payment for right of way
**cheminium, cheminum, cheminus**, road, way
CHEP *see* **chippa**
CHERCHESED *see* **cyricsceattum**
**cherna, churna**, CHERNE, milk-churn
CHESHOWS *see* CHASHOWS
**chevagium, chivagium**, 'head-penny', manorial rent
CHEVEN, chub, chevin
**cheverellus**, kid
**chevero, cheveronus**, CHEVERONE, piece of timber; rafter
CHEVERUND *see* **severunda**
**chevetta, chevicium**, CHEVETT, headland
CHEYNEHOKE, chain-hook, drail
**chiminagium** *see* **cheminagium**
**chipclutum, kipclutum**, iron patch on ploughshare
**chippa**, CHEP, CHIPPE, 'chep', share-beam of plough

**chippo**, to fit a 'chep'
CHIRCHELOFE, bread given to church at Christmas; *cf.* HOLILOFE
**chirotheca, cerotheca, cirotheca, sarateca**, glove (worn by labourer)
**chivagium** *see* **chevagium**
**cholettus**, CHOLETT, small cabbage
**chopina**, half-pint, chopin
**chorda, corda**, cord, rope
**churna** *see* **cherna**
**cibaria**, victuals; food allowance
**cibus**, food
**cicera** *see* **sicera**
**cignus** *see* **cycnus**
**cilicium**, haircloth, sacking
**ciminum** *see* **cyminum**
**cinapium** *see* **sinapium**
**cindula** *see* **scindula**
**cingulum**, SYNGUL, girdle, girth (of harness)
**cinis**, ash, potash; *see also* **dies**
**ciphus** *see* **scyphus**
**cippagium, seppagium**, (right to take) tree-stumps
**cippus, sippus**, tree stump, stock
**circulator**, hooper
**circulo**, to hoop (cask or wheel)
**circulus, cerculus, sirclum**, hoop; loop; ring
**ciresum** *see* **cerasum**
**cirotheca** *see* **chirotheca**
**cirpus** *see* **scirpus**
**cisara** *see* **sicera**
**cista**, chest, box
**civera, severa**, barrow; **c. rotabilis, rotata**, wheel-barrow
**clades, claia** *see* **cleta**
**clapera** CLAPER, CLAPPER, rabbit-hole; hutch
**clapsa, clapsum, claspum**, clasp, fastening (on harrow)
**claudo**, to close; to enclose; **clausum**, close, enclosed yard; *see also* **hercia; tabula**
**claustro**, to close; to enclose
**clausum** *see* **claudo**
**clausura, claustura, clostria**, fence, or material for fence
**clava**, club, mace
**clavis**, key

**clavus**, nail; **c. gariophili**, clove
**cleia** *see* **cleta**
CLEPPEL *see* **clipella**
**cleta, clades, claia, cleia, cleya**, hurdle; *cf.* **cratis**
CLEVIS, CLEVIE, CLEVYE, piece of iron connecting ploughbeam to draught animal
**cleya** *see* **cleta**
**clippella, kippella**, CLEPPEL, CLIPPEL, wheel-clamp
**cloaca**, drain, sewer
**cloca**, (riding) cloak
CLOSHE BOULLS, game (often prohibited) resembling croquet
**clostria** *see* **clausura**
CLOTHIER, cloth worker
CLOUT *see* CLUT
CLOVESTAPLE, split bar, used in framed building
**cloverus**, CLOVIER, cleaver, splitter
CLUT, CLOUT, CLOWT, metal plate for repairing implement
CLUTNAYL, CLOUTNAYL, cloutnail
**cluto**, to patch
**cochlear, cocliar**, spoon
COCKIL, corn cockle
**cocreo, cocrio**, (?) to pipe (horse-traces) with leather
**coffinus** *see* **cophinus**
**cogga**, cog
**coggatio**, fitting of cogs
**coggo**, to fit w. cogs
**cokettus 1**, seal; certificate
**cokettus 2**, loaf of bread
**cokettus 3**, hay-cock
COKSHOTE, cockshoot (trap for woodcocks)
**colara** *see* **collare**
**coleicium, culacium**, 'cullis', lean-to building
COLIER, COLYER, charcoal burner
**collare, colara, colera, collaria, collarium**, horse-collar
COLETTEFORKE, (?) cabbage fork
**colliculus**, hillock
**collis**, hill; **c. talpae**, mole-hill
**collistrigium**, pillory
COLQUENCHE, charcoal ash (still combustible)

**columba**, dove; pigeon
**columbare, columbarium**, dovecot
**columbella**, young dove
COLYER *see* COLIER
COMBE *see* **cumba**
**comburo**, to burn
**combustio**, burning; **c. rotulorum**, burning of records during Peasants' Revolt
**comedo**, to eat
COMELYING *see* **cumelingus**
**comestio**, eating; food
**comestus**, food; **c. porcorum**, pig-swill
**communa, communia** (right of) common; common land; **c. pasturae**, common of pasture
**communico, communo**, to graze beasts on common
**communis**, common, communal; *see also* **via**
**communitas pastoragii**, common of pasture
**companagium, comparnagium, cumpanagium**, relish eaten w. bread
**compes**, fetter; *see also* **par**
**composto, composturo**, to manure
**compostum**, manure
**computus, compotus**, financial account
**concelamentum, consulamentum**, concealment
**concicularium** *see* **cunicularium**
**conculco**, to trample down; to crush
**conduco**, to lead (water etc.); to hire
**congrego**, to bring together; to gather in
**constamentum, custamentum**, cost, outlay
**constantiae** (*pl.*) expenses
**constringo**, to confine; to distrain
**consuetudinarius**, customary
**consuetudo**, custom, usage; customary payment
**consulamentum** *see* **concelamentum**
**conventio**, agreement; contract
**convillanus**, fellow villein
CONYWER, CONYNGER *see* **cuningaria**
COOKYNGSTOLE, ducking-stool
**cooperio, coperio**, to cover
**coopertio, coopertorium, coopertura**, roofing; covering; thatching

**coopertor**, thatcher
COOPESELVER, (feudal) payment
COORIER, currier; C. **albus**, whitetawyer; *cf.* **correator**
COPE, coop; C. **ad faldam**, shepherd's hut
COPELE *see* **copula**
**coperio** *see* **cooperio**
**copero, coperonus**, COPERONES, loppings, 'lop and top'
**cophinus, coffinus**, box, chest
**copia**, copy (of court roll, w. ref. to copyhold); **teneo per copiam**, to hold by copy
**copicia, copicium, copusa**, coppice wood
**coppa**, stook, 'cop'
COPPEDLOVES, (?) cottage loaves
**coppo**, to stook
**copula**, COPELE, couple (of tie-beams)
**copusa** *see* **copicia**
**coquina**, kitchen; food (allowance)
**corallum** *see* **curallum**
**corbellus, corbella, corbilus**, basket
**corda** *see* **chorda**
CORDEWANER, cordwainer, leather-worker
**corium, coreum**, hide, leather
**cornarium, cornera**, corner
**cornemusa**, CORNMUSE, (bag) pipe
CORNERNAYLL, corner-nail
CORNERTEILL, corner-tile
**cornu**, horn (of animal); hunting- or moot- horn
**coronator 1**, coroner
**coronator 2** *see* **cronarius**
**corporo, corpero**, to fit (plough) w. new body
**correator**, currier; *cf.* COORIER
**corredium, corrodium**, corrody, food allowance
**correo**, to curry, to dress (leather)
**corrigia**, measuring-strap; thong
**cort-** *see also* **curt-**
**cortillagium, curtillagium**, curtilage, yard
**cortillarius, curtillarius**, servant in charge of curtilage
**corylus, corulus**, hazel
**cos**, whetstone
**costa**, rib; bar; ridge; bank (of ditch); side (of building)

**costagium, custagium**, cost, outlay
**costrellus, costerellus**, keg; ladle
**cota**, coat
**cotagium**, cottage (tenure or building)
**cotaria, coteria**, cottage
**cotarius, coterellus**, cottar, cottager
**cotlanda**, COTLOND, COTMANNELOND, cotland, cottar's holding
**cotmannus**, COTMAN, cotman, cottar
**coto**, COTOUN, cotton; *see also* **candela**
**cotsetla, cotsetlus, cotsetus**, cottar; cotland
**courtus** *see* **curtus**
COWLEAZE, cow-pasture
CRABBETREW, CRABBETRIWE, CRABTRE, crab-apple tree
**cracha, crechia**, CRACHE, crib, manger
**craculus** *see* **graculus**
CRADYL, feeding rack for sheep
CRANAGE, cranage, hoisting fee
**crassetum, crassatum**, cresset, lamp
**crastello** *see* **cristo**
**craticula**, griddle
**cratis, cratea**, hurdle; *cf.* **cleta**
CRAWE, 'crow', crowbar
**creca, crekum**, CREKE, CRIKE, creek
**crechia** *see* **cracha**
**crementum**, increment; increase of rent
**cresco**, to grow; **crescentia**, growing crops
**cresta** *see* **crista**
**cretina**, flood
**cribro**, to sift
**cribrum**, sieve
CRIKE *see* **creca**
**crinetum**, (?) haircord
**crinis**, hair
CRIPS, (?) crepe
**crispus**, curly, frizzy
**crista, cresta**, crest; ridge; embankment
CRISTEMASSECUSTOM, due of one penny paid at Christmas; *cf.* OFFRYNGPANY
**cristo, crastello, cristello**, to cap, to cope (bank or wall)
**crocus 1**, saffron
**crocus 2**, crook, hook; part of plough-harness
**crofta, croftum, crufta**, croft, enclosed field

**croftarius, croftmannus**, crofter
CROKKE, crock, jar
CROMB, CROME, hooked stick
**cronardus, cronatus**, CRONE, crone, worn-out animal
**cronarius, coronator, cronator**, carcase inspector
**crono**, to cull crones
**croppa**, crop, harvest
**croppus**, crop and lop (of trees); *see also* LOPPES AND CROPPES
CROPPYNGE, cropping, clipping
CROSBERD, CROYSBRED, crossboard on shafts of cart
CROTCH, CRUTCH, tie-beam; piece of forked timber
CROUCH, CRUCHE, (wayside) cross
CROUCHEWERE, kind of weir; *cf.* **wera**
CROUDEWAIN, CROWDWAYNE, hand-cart
CROYSBRED *see* CROSBERD
CRUCHE *see* CROUCH
CRUDDISMILL, CRUDEMELNE, curd-mill (for cheese)
**crudus**, raw, untreated
**crufta** *see* **crofta**
CRUTCH *see* CROTCH
**crux**, cross
**cubatio**, setting in place; (of eggs) incubation
**cubo, cumbo**, to lie down; **cumbans**, broody (hen); **cumbans et levans**, permanently resident
**culacium** *see* **coleicium**
**cultellus**, knife
**culter, cultrium**, knife; plough-coulter
**cultura**, cultivation; strip of ploughland or meadow
**cumba**, COOMBE, coomb, dry measure, us. four bushels; *see also* LODCUMB
**cumbo** *see* **cubo**
**cumelina, kemelinus**, tub, vat
**cumelingus**, COMELYNG, CUMELYNG, newcomer; stray animal; CUMELINGESLOND, land held by C.
**cuminum** *see* **cyminum**
**cumpanagium** *see* **companagium**

**cumulatio**, heaping
**cumulo**, to heap, to pile up (esp. grain); *see also* **mensura**
**cumulus**, heap; heaped measure; stook
**cunae, cuna**, cradle
**cuneus, cunea**, wedge
**cunicularium, concicularium**, rabbit-warren
**cuniculus**, rabbit
**cuningaria, cuningera**, CONYNGER, CONYWER, rabbit warren
**cupa**, tub, vat, bowl; *cf.* **cuppa; cuva**
**cuparius**, CUPER, cooper
**cuppa**, cup; *cf.* **cupa; cuva**
**curallum, corallum, curaltum, scurallum**, chaff
**curba**, curved piece of wood; (?) felloe
**curbo**, to fit w. (?) felloes
**curia**, court, courtyard; manorial court
**curtill-** *see* **cortill-**
**curtus, cortus, courtus**, short; *see also* **carretta**
**custagium** *see* **costagium**
**custamentum** *see* **constamentum**
**custodia**, (safe) keeping
**custodio**, to keep, to guard; **custodiens**, keeper; watchman
CUSTOMPOTTES, ale-toll at brewing-time; *cf.* **olla; tolsester**
**custos**, keeper; **c. autumni, autumnalis**, keeper of harvest by-laws
**custuma**, customary service or payment
**custumarius**, customary, given as payment; *see also* **olla**
CUSTUMBEDREPPE, harvest boonwork
**cuva**, tub; *cf.* **cupa; cuppa**
**cuvarius**, cooper
**cycnettus, cigniculus**, SIGNET, cygnet
**cycnus, cignus, cygnus**, swan
**cyminum, ciminum, cuminum**, cumin (esp. as rent)
**cyphus** *see* **scyphus**
CYRICSCEATTUM, CHERCHESED, customary payment (sometimes of wheat), to parish church, 'church-scot'
**cysera** *see* **sicera**

# D

**dabbo** *see* **daubo**
**dagardus, dagardum, dagardium, dagg-**, dagger
DAGSHOU, DAKESHO, (?) part of mill mechanism
**daia, daya,** DAYE, DEYE, dairy-maid
**daieria, dayeria**, dairy, dairy-farm
**daiwerca, daiwercata, daywerca**, DAUWARK, DEYWERKE, 'day-work', measure of land; *cf.* **diaeta**
**daiwerclanda**, 'day-work', measure of land
DAKESHOU *see* DAGSHOU
**dala, dola, dolla,** DOOLE, 'dole', piece of land esp. meadow
**daubatio, daubatura, daubitura**, plastering
**daubator**, plasterer
**dauberia, daubura**, plastering
**daubo, dabbo, dawbo, dobo**, to daub, to plaster
DAUWARK *see* **daiwerca**
**dawbo** *see* **daubo**
**day-** *see* **dai-**
DAYE *see* **daia**
**dealbator**, whitewasher; whitetawyer
**dealbo**, to whitewash; to taw
**decado**, to fall into ruin; **decasus** (*p.p.*), ruined
**decasus**, ruin; loss
**decena, decenna, discena**, tithing, frankpledge group
**decenarius**, tithingman, headborough
**decimalis**, for tithes, tithable; *see also* **grangia; horreum**
**decimator**, tithe collector
**decimo**, to pay tithe
**decimus**, tenth; **decima, decimae**, tithe(s)
**decius**, dice (*us. pl.*)
**decorticatio**, stripping bark
**decortico**, to strip bark, to peel; *cf.* **discorio**

**defendo**, to 'defend' land as game preserve, or by enclosure; **d. me**, to be assessed at (of land); **defensus**, enclosed (of land)
**defensio**, 'defence'; seasonal enclosure (of land)
**defrondo**, to lop, to prune
**delacero, delassero**, to tear apart, wreck
**denariata**, penny-worth (esp. of land)
**denarius**, penny; **d. cerae**, waxshot
**depasturo**, to pasture (animals); to depasture
**depenno**, to pluck (feathers from bird)
**derelinquo**, to abandon; to desert; **derelictus**, derelict
**desicco**, to dry up, to drain; (of cow) to go dry
**devillo**, to abscond; to leave the 'vill'
DEWRAKE, rake for grass or stubble
DEWRODE, service (?) of raking
DEYE *see* **daia**
DEYEHOUS, dairy
DEYWERKE *see* **daiwerca**
**diaeta**, working day; measure of land; *cf.* **daiwerca**
**dies**, day; working day; **d. cinerum**, Ash Wednesday; **d. Palmarum**, Palm Sunday; **d. Parasceve**, Good Friday; **d. Pasche**, Easter Day; *see also* **hockdies**
**dignerium** *see* **dinarium**
**dilato**, to spread, to scatter
**dimitto**, to grant (land, esp. **ad firmam**)
**dinarium, dignerium**, dinner
**diplois, duplodium**, doublet
**discena** *see* **decena**
**discooperio**, to uncover, to unroof
**discorio**, to bark (trees); *cf.* **decortico**
**discus**, dish, measure (of grain, salt etc.)
**dishereditatio**, dispossession
**disheredito**, to dispossess
**disjungatio**, unyoking
**disjungo**, to unyoke
**dispergo**, to spread (hay or manure); to scatter
**districtio**, distraint

**distringo**, to distrain
**divido**, to divide
**divisa, divisum**, boundary (mark)
**dobo** *see* **daubo**
**dodda**, measure of grain
DOGGE, clamp (used in building)
**dola, dolla** *see* **dala**
**dolatio**, hewing
**doleum, dolium**, cask
**dolo**, to hew
**domus**, house, building; **d. casei**, cheese house, *cf.* CHASHOWS;
    **d. forinseca**, outbuilding;
    **d. gallinacea**, henhouse; **d. ignea**, kiln; **d. lavendrie**, laundry;
    **d. placitorum**, courthouse;
    **d. porcorum**, pigsty; **d. torcularis**, press house
DONGECOPPE *see* DUNGCOPPE
DOOLE *see* **dala**
DORELEGGE, (?) door jamb
DORENAIL, stud nail for door
DORERIDE, strap for door hinge
DORESTALL, door jamb
**dormio**, to sleep; *see also* **tabula**
**dorsarium**, back-band (for carthorse)
**dosinus**, †**doscus, doseus, dosius**, dun, grey
**douella**, DOULE, DUUELE, dowel (for cart-wheel)
**douellegia**, DOWLEGGE, DULLEGE, 'duledge', dowel
**douello**, to fit w. dowels

DOYLLOUR, (?) doweller
**dragetum**, DRACHAT, 'dredge', mixed corn (oats and barley)
**dragla** *see* **draila**
**draia, dragia, draya**, DRAIE, DRAYE, dray; dray-load
**draila, dragla, drailla, drayla, dreyla**, DRAIL(E), DRAYL, drail, notched iron on plough-beam
**drasca, drascum, drassum**, malt-dregs, used as animal food
DRAUGHT **pons**, drawbridge
**draya**, DRAYE *see* **draia**
**drayla** *see* **draila**
**dresso**, to dress; to set in order
**dressorium**, DRESSORE, dresser, serving-table
**dreyla** *see* **draila**
**drova**, driveway
DUE, (?) vessel or implement in dairy
DUFFUS, dovehouse
DULLEGE *see* **douellegia**
**dumetum, dumetus**, brier patch; thorn bush
DUNGCOPPE, DONGECOPPE, DUNGCOUPE, dung-cart or tub
DUNGHOPE, dung-hoop
DUNGKORVE, dung-basket
DUNKHOK, dung-hook
**duplex**, double; *see* **hercia**
**duplodium** *see* **diplois**
**durus**, hard; *see also* **bladum**
DUUELE *see* **douella**

# E

**ebdomas, ebdomada** *see* **hebdomas**
EBWERE, weir for trapping fish at ebb-tide; *cf.* **wera**
**edera 1**, EDDER, edder, osier, or other flexible wood (used for binding fences)
**edera 2** *see* **hedera**
**edero**, to interlace w. osiers
**edis** *see* **haedus**
**edulium**, food
**effortio, affortio**, to reinforce, to strengthen

**effraia, affraia**, affray
**effundo**, to pour out; **effusus**, scattered
**elargatio**, widening, enlargement
**elargo, elargeo, elargio**, to widen, to extend
**elongo**, to remove; **e. me**, to depart, to absent oneself, esp. from custody or service
**emendatio**, repairing
**emendo**, to repair
**emergo**, to submerge, to flood

**emundo**, to clean; to clear (ground)
ENDWEY, cul-de-sac
**eneus** *see* **aeneus**
**ensiculus, hensiclius**, small sword
**ensis**, sword
ENYNCORN *see* YENYNGCORN
**equa**, mare
**equestris** *see* **via**
**equinus**, of, for, or operated by horse; *see also* **falda; panis**
**equitium**, stud
**equitius**, operated by horse; *see also* **molendinum**
**equito**, to ride
**equo** *see* **aequo**
**equus**, horse; **e. carrettae**, cart-horse; *see also* **molendinum; passus**
**eradico**, to uproot, to remove
ERDLYNG *see* YERDLING
EREBREDE, front board of cart
**erigo**, to raise; to set up; to lift; **erectus**, upright; lifted up
ERYERE, (?) ear-iron (of plough)
**es** *see* **aes**
**escaeta, kaeta**, escheat; wood fallen from tree; material from dismantled building
**escaldaria, ascaldaria, schaldaria**, scalding
**escaldo**, to scald
**eschina**, chine, backbone
**escuro, exscuro**, to scour, to clean
**esperduta, asperduta, experduta**, metal rod, crowbar
**essarta, assarta, essartum, exsarta, sarta**, assart, area reclaimed from woodland
**essarto**, to assart
**essewera**, drain; ditch; sewer
**estaka** *see* **stica**
**estas** *see* **aestas**
ESTERCHBORD, ESTRICHBORD, timber from the Baltic
**estivalis** *see* **aestivalis**
**estivatio** *see* **aestivatio**
**estoppo** *see* **stuppo**
**estoverium**, estovers, allowance of food or wood

**estraia, extrahura, extraura, straia**, stray (animal)
**estrappamentum**, EXTRIPMENT, stripping
**estrepum, strippum**, ESTREP, stripping
ESTRICHBORD, *see* ESTERCHBORD
ESTRICKE, 'strike', strickle, implement for levelling off a measure of grain; *cf.* **strica**
**estuffum** *see* **stuffum**
**estupamentum**, (?) mill-dam; *cf.* **stuppo**
**etas** *see* **aetas**
ETUE *see* STUE
EUDEVY, eau-de-vie, brandy
**evagino**, to unsheath
**evello**, to tear up; to pull down
EVESDROPE, (?) area receiving water from house eaves
EVESLATHE, timber for eaves
EWYN, (*pl.*), ewes
**exalto**, to build up, to make higher
**examen**, swarm; **e. apium**, swarm of bees
**excarrio**, to carry away
**exclusa**, sluice, lock, esp. of mill
**excorio**, to skin, to flay; to shell (peas)
**excultibilis**, cultivable
**exenium, exhennium**, offering, customary payment
**exitus**, issue; profits; produce
EXNAIL *see* AXNAIL
**exorbito**, to deviate, to stray
**expando**, to spread
**experduta** *see* **esperduta**
**explectum, explectium, expletia**, revenue
**expurgatorium**, drain
**exsarta** *see* **essarta**
**exscuro** *see* **escuro**
**estraho**, to pull or draw out; **e. me**, to abscond (from manor)
**extrahura** *see* **estraia**
**extraneus**, from outside, foreign
**extraura** *see* **estraia**
EXTRIPMENT *see* **estreppamentum**
EYREN (*pl.*), eggs

# F

**faba**, bean
**faber**, blacksmith; **fabra**, (?) female blacksmith
**fabrica**, smithy; workshop
**fabrico**, to forge, to work in iron
**facio**, to do; to make
**factus 1** (*p.p.*), done, made
**factus 2**, *see* **vatta**
**fagoto**, to make into faggots
**fagotus, fagattus**, faggot
**fagus**, beech tree
**faissus, fessus**, bundle
**falcastrum**, billhook
**falcatio, falcatura**, mowing
**falcator**, mower
**falco 1**, to mow
**falco 2**, falcon
**falda**, fold, enclosure for animals; *see also* COPE; **f. equina**, stud-fold for horses
**faldagium**, 'faldsoke', obligation to fold livestock on lord's land, or payment in lieu; *cf.* **faldsoca**
FALDBOTE, FOLDBOTE, right to take wood for folds
**faldicium**, manuring (land, by folding livestock)
**faldo**, to fold (land or animals)
FALDSELVER, 'faldsoke'
**faldsoca**, 'faldsoke'; *cf.* **faldagium**
**fallaratus**, adorned; *cf.* **phalera**
FALLYNGEGATE, falling sluice
**falx**, sickle; scythe; **f. gemina**, double scythe
**familia**, household
**famulus, famula**, servant
FANFULL, as much as a (winnowing) fan can contain
**fanna** *see* **fennum**; **vanna**
FANNEMAKER, maker of (winnowing) fans
**fanno** *see* **vanno**
FARCYNE, FERSYOUN, farcy, disease of horses
FARE, litter of pigs
FAREBOT, (?) ferry boat
**fario**, (?) to treat sick animals
FARTHENDALE, FARTHING, fourth part; quarter (? dry measure); *cf.* **ferthendella**
**fasciculus, fascicula, fessiclus**, faggot
**fattum**, FATTE, *see* **vatta**
†**faviculus** *see* **paviculus**
FAYSONE, pheasant
**felea, velia**, VELIE, VELY, felloe of wheel
**felgera, feugera, fougera, fugera**, bracken
**femerellus** *see* **fumerellus**
**femicium** *see* **fimitium**
**fenatio**, haymaking
**fencibiliter**, †**fensibiliter**, fencibly, capable of being defended
**fenestra**, window
**fenile**, hay loft; haystack
**fennum, fanna**, VANNE, fen, marsh
**feno**, to mow; to make hay
†**fensibiliter** *see* **fencibiliter**
**fensura**, fence, fencing
**fenum**, hay; **f. salsum**, salt hay
**feodum**, fee, fief; estate
**fera**, beast, esp. deer
FERBOTE *see* **firbota**
**ferculum**, dish (of food); morsel
FERDEKYN, firkin
**feretrum**, bier
**ferlingata, ferlingatum**, 'ferling', quarter-virgate; *cf.* **quarentena**
FERLINGLOND, FERTHYNGLOND, 'ferling'
**ferlingus**, VERDLINGGE, VERDLYNGE, 'ferling'; *cf.* YERDLING
FERMELCH (*adj.*) 'farm-milk' (of cows leased out while in milk)
FERMELOND, land held on lease
FERMEREYE, farmery, farm buildings
†**ferndiclus** *see* **ferthendella**
FEROUR, smith; farrier
**ferramentum**, ironwork; horseshoe
**ferrarium**, smithy; farriery

**ferratio**, fitting w. ironwork
**ferro**, to shoe (horse); to fit w. iron; *see also* **biga**; **carretta**
**ferrum, ferrus**, iron (object, esp. horseshoe); **f. pedale**, plough-foot; *see also* STIROP
FERSYOUN *see* FARCYNE
**ferthendella**, †**ferndiclus, ferthendellus**, 'fardel' (of land), quarter-virgate; (of dry goods) quarter barrel; *cf.* FARTHENDALE
FERTHYNGLOND *see* FERLINGLOND
**fesianus** *see* **phasianus**
**fessiclus** *see* **fasciculus**
**fessus 1**, *see* **faissus**
**fessus 2**, weary, worn out
**festum 1**, ridge of roof
**festum 2**, feast-day
**fetherborda, fithelborda**, FETHERBORD, VYTHELBORD, board w. one feathered edge
FETLOCK, VETTERLOK, VTERLOK, fetterlock
**feugera** *see* **felgera**
**fidelitas**, (oath of) fealty; feudal service
**figulus**, potter
FILBERD, filbert, hazelnut
**filix**, fern, bracken
**filo**, to spin
**filum, philum**, thread
**fimaculum, fimale, fimarium**, dunghill
**fimitium, femicium**, manuring
**fimo**, to manure
**fimum, fimus**, dung
**findo**, to split; to cleave
**finis**, end; boundary; fine (esp. w. ref. to conveyance)
**firbota**, FERBOTE, right to gather firewood
**firgia**, fetter
**firma**, rent; *see also* **dimitto**; **jaceo**; **f. alba**, blanch farm, rent paid in silver; *cf.* BLANCHEFERME
**firmaculum**, buckle
**firmale**, list of farm tenements
**firmarius**, farmer
**firmo 1**, to make firm, to secure
**firmo 2**, to farm, to pay rent
**fistula**, pipe

**fithelborda** *see* **fetherborda**
**fixator**, pitcher (farm worker)
**flado, flato**, 'flawn', pudding
**flagello**, to thresh
**flaka**, FLAKE, FLAKKE, wattled hurdle
**flato** *see* **flado**
FLAUNDRISSHTYLE, Flemish tile
**flecca, flecta**, arrow-shaft
FLEET *see* **fleta**
FLEMDICHE, millstream
**fleta, fletum**, FLEET, FLETTE, fleet, creek, channel
FLEX, flax
**flocca, flokka**, flock (of sheep or other animals)
**floccus**, FLOKKE, flock, wool refuse
**flodum, flotta**, (mill) pond; *see also* **porta**
FLODWERE, weir (? to catch fish at flood-tide); *cf.* **wera**
**flokka** *see* **flocca**
FLOKKE *see* **floccus**
FLOODYATE *see* FLOTGATE
**floridum** *see* **Pascha**
FLOTGATE, FLOODYATE, flood-gate; sluice-gate; *cf.* **porta**
**flotta** *see* **flodum**
**fluvialis**, of or for water; *see also* **porta**
**fluvius, fluvia**, river; flow of water; flood
**focagium**, 'focage, hearth-tax
**focalis**, for a hearth; **focale**, fuel
**focarius**, related to a hearth
**focus**, hearth
FODDERBYNNE, bin for fodder
FODDERCORN, FODDERKORN, payment of fodder
FODDERSELVER, payment in lieu of foddercorn
**fodio**, to dig
**foditio**, digging
FOLDBOTE *see* FALDBOTE
FOLDHURDLE, hurdle for making sheep-fold
**folerectus** *see* **fullericius**
**folium**, leaf; **f. hostiae**, leaf of door
**fons**, well
**fonticulus**, small well
**foracra**, foreacre, headland

**foragium**, forage, fodder, straw
**foramen**, hole, opening
**foranus, foraneus**, external, outside (the manor)
FORBAY, front bay; *cf.* **baia 1**
**forceps, forpex**, (pair of) tongs; shears
**forcerum, forserium**, strong box
†FORDERMALT, (?) service of making malt
**fordus, furdus**, ford
**foredrova, fordruvia**, FORDROVE, FOREDROVE, foredrove, gift of horse or cow led before funeral cortege
**forefactura** *see* **forisfactura**
FOREFURROWED, FORISFURWYD, of the first ploughing
FOREGRIST, (?) right of precedence in grinding corn
**forelandus**, FORELAND, FORLAND, outlying land, leased on special terms
**forelina**, FORELYNE, FORLYNE, rein of cart
**forera, forherda, forraria**, headland
**forestallator**, forestaller
**forestallo, forstallo**, to forestall (the market)
**forestallum**, FORESTALL, FORSTOLL, front part of cart
**forettus** *see* **furettus**
FOREYERD, yard in front of building
**forgea, forgia, forgium**, forge, smithy
**forherda** *see* **forera**
**forinsecus**, outer, outlying, outside; *see also* **boscus; domus**
**forisfacio**, to forfeit
**forisfactura, forefactura**, forfeiture
FORISFURWYD *see* FOREFURROWED
FORKESYLVER, payment in lieu of forking service
FORLAND *see* **forelandus**
FORLYNNE *see* **forelina**
**forma**, mould (esp. for cheese); form, bench
FORMANLONDE, land held by **formannus**
**formannus**, tenant of part virgate
**formella, formula**, small mould
**fornacium, fornasium, fornax, fornitium, fornus**, oven (esp. baker's); kiln
**forno, fornio, furnio**, to bake
**forpex** *see* **forceps**

**forraria** *see* **forera**
**forserium** *see* **forcerum**
**forstallo** *see* **forestallo**
FORSTOLL *see* **forestallum**
**fortia**, force; **frisca f.**, 'fresh force', forcible dispossession
**fossa, fosseia**, ditch
**fossator**, ditcher
**fosso**, to dig; to raise (bank); **fossatum**, ditch
**fotaverium**, FOTAVER, carrying service on foot
**fothera**, FOTHER, fother, measure of weight
FOTNAYL, nail used in tiling
**fougera** *see* **felgera**
FOURWENT, FURWANTES, cross-roads; *cf.* WENT
**fovea**, pit, ditch
**fraccinus** *see* **fraxinus**
**fractio**, breaking (into)
**fraellus, fraella**, FRAYELLE, frail, fruit basket
FRAMPOLE, (?) frame-pole
**franclingus**, FRANKLYN, franklin, freeholder
**frango**, to break (into)
FRANKLYN *see* **franclingus**
**fraxinus, fraccinus, fraxina**, ash-tree
FRAYELLE *see* **fraellus**
**frenum**, bridle, bit
**fretta, frettum**, hoop for tub or wheel
**fretto**, to bind w. hoop
**friscus**, fresh; not salt; *see also* **fortia; terra**
**frithsocna**, FRITHSOKEN, FRYTHSOKNE, view of frankpledge
**frixo**, to fry
**frixorium**, frying pan
**frucisium** *see* **frussatum**
**fructus**, yield, crop; fruit
**frues** (*pl.*), rooks
**fruges** *see* **frux**
**frumentum**, crop, esp. wheat
**frunio**, to tan
**frunitor**, tanner
**frussatum, frucisium**, newly broken land
**frustum, frustrum**, scrap, lump; piece (of land)

**frutectum, frutetum**, thicket
**frux** (*us. pl.* **fruges**), crops, esp. corn
FRYTHSOKNE *see* **frithsocna**
**fucellus** *see* **fusillus**
**fugalis** *see* **via**
**fugatio**, driving (away)
**fugator**, driver; **f. carrucae**, ploughman
**fugera** *see* **felgera**
**fugo**, to drive
**fugio**, to flee (from)
FULFILLINGDAY, view of frankpledge, day when those newly admitted were drafted into tithings
**fullericius, folerectus, fullonicus**, of or for fulling; *see also* **molendinum**
**fullo 1**, to full (cloth); to tread down (hay)
**fullo 2**, fuller
FULLYNGSTOK, wooden mallet on fulling-mill
**fumerellum, femerellus, fumerellus**, smoke-vent
**fumus**, smoke

**funda**, strap; sling
**fundamentum**, base; foundation
**fundo**, to found, to establish
**fundus**, base, bottom
**fungia, funga**, stockfish
**furca**, fork, pitchfork; forked timber; (*pl.*) gallows
**furcarius, furcator**, pitcher, loader
**furdus** *see* **fordus**
**furfur**, bran
**furnio** *see* **forno**
**furettus, forettus**, ferret
**furro**, to line w. fur; to plate w. metal
FURWANTES *see* FOURWENT
**fusella** *see* **fussella**
**fusillus, fucellus**, spindle
**fussella, fusela**, 'stick' (of eels)
**fustanium, fustianum**, fustian
**fustis**, stick, cudgel
**fustum**, stick; tree-trunk
FYNSPARRIS, VINSPARRES, VYNSPARRIS, projecting rafters of barn
FYRESTOKE, fireplace

# G

GABLE *see* GAVEL
GABLECOPEL, beam across gable; (?) gable-fork
**gabulum 1, gablum**, tribute, gavel, rent; *cf.* GAVEL
**gabulum 2**, gable of building
**gaddum**, GAD, GADDE, 'gad', metal rod, goad
GADERCART, cart for gathering (hay)
**gainagium** *see* **wainagium**
**galerus, galerum, gallirum**, hat
**gallina**, hen
**gallinaceus** *see* **domus**
**gallinaria**, hen-house
**gallirum** *see* **galerus**
**gallus**, cock
**galo, galona, galonus**, gallon; measure
GALUNSELVER, (?) brewing payment
**ganga**, set of cogs, spindles
**gannoco, gannoko**, to sell ale
GANNOKER, GANOKER, alehouse-keeper

**gara** *see* **gora**
**garba**, sheaf
**garcia** *see* **gercia**
**garcio**, boy; servant
**gardenarius** *see* **gardinarius**
**garderius** *see* **wardarius**
**garderoba**, privy
**gardinarius, gardenarius**, gardener
**gardinum**, garden; orchard
**garenna** *see* **warenna**
**gariofilum, gariophilum,** *see* **caryophyllum**
**garita**, GARYTE, watch-tower; garret
GARLED, GARLYD, 'garled', spotted, speckled
**garnaria** *see* **granaria**
GARNET *see* **carnettus**
**garrulatrix, garulatrix**, scold; chatterbox
**garsacra**, GRASAKER, ploughing service in return for grazing rights

# GARSAVESE

GARSAVESE *see* GRASAVESE
GARSTAPLE, GASTAPLE, (?) gate post
**garulatrix** *see* **garrulatrix**
**garuna** *see* **warenna**
GARYTE *see* **garita**
GASTAPLE *see* GARSTAPLE
**gastellum** *see* **wastellus**
**gata**, bowl, trough
GATEPANY, toll paid by tenant at lord's gate
**gatta**, gap; gateway
GAVEL, GABLE, payment, service, 'gavel'; G. **aro** to perform ploughing service; *cf.* **gabulum 1**
GAVELACRE, reaping service
GAVELCORN, corn tribute; *cf.* GAVELESED
GAVELERTHE, GAVELHERTHE, GAVELYRTHE, GAWELHERTHE, ploughing service
GAVELESED, threshing service; (?) corn tribute; *cf.* GAVELCORN
**gavelmannus**, tenant paying 'gavel'
GAVELSELVER, payment in lieu of 'gavel'
GAWELHERTHE *see* GAVELERTHE
**gaynagium** *see* **wainagium**
**gayneria** *see* **waineria**
GEG, worn out
GEMEL, GEMEW, hinge
**gemellus**, twin; forming a pair
**geminus**, twin; *see also* **falx**
**genesta, genectum, genesteium, jenectum**, broom, furze
**genesteia, geneteia**, broom-land
**gent-** *see* **jent-**
**gercia, garcia, jercia**, gimmer, maiden ewe
GERDYLL, griddle
**germen, jerma**, shoot, young growth
GERNETT *see* **carnettus**
**gersuma**, premium, payment; merchet
**gersumo**, to pay premium
GERTCHEZ, GERTES, girths (for horse)
GERTHECLOTH, girth-webbing
GILT, YELT, young sow, gilt
**gingiber** *see* **zinziber**
**giro** *see* **gyro**
**gista**, GISTE, JYSTE, length of timber; joint; joist
**glaiva, gleyvus**, bill, cutting tool or weapon
GLANDRES, glanders, disease of horses

**glans**, acorn; mast, pig-food
GLAS, glass
**gleyvus** *see* **glaiva**
**globus**, ball; *see also* **ludus**
**gogeum, gojo**, GOGEON, GOJOUN, gudgeon
**golfa** *see* **gulfus**
GOLVERE, corn-stacker
**golvo**, to stack corn
**gora, gara**, GORE, gore, triangular piece of land; *cf.* **gyro**
**gorgius** *see* **gurges**
GOSEPANNE, (?) pan w. gooseneck handle
**goterum** *see* **guttura**
**graculus, craculus**, jackdaw; rook
**gramen**, grass
**granaria, garnaria, granarium**, granary
**granarius, granatarius, granetarius**, granary-keeper, granger
**grando**, hail
**granetarius** *see* **granarius**
**grangia**, barn, grange; **g. decimalis**, tithe barn
**grangiarius**, granger
**granum**, grain
GRASAKER *see* **garsacra**
GRASAVESE, GARSAVESE, pannage; *cf.* **avesagium**
**grata**, GRATE, grating; (?) grater
**grava, grova**, grove
**gravera**, sandpit
**grena**, (village) green
GRESELDE *see* **grisellus**
**grex**, flock, herd; swarm
**gripa 1**, ditch; drain
**gripa 2** *see* **gropa**
GRIPE *see* **gropa**
**grippella**, small ditch
**grisellus**, GRISEL, GRESELDE, grey
**grisus**, grey
**grondsella** *see* **groundsella**
**gropa, gripa**, GRIPE, GROPE, GROPPE, GRYPE, 'grope', iron plate reinforcing cart-wheel
GROPENAILE, GRYPNAIL, nail for fixing 'grope'
**grossus**, large, stout; (of livestock) cattle; (of fleece), coarse; **in grosso**, altogether, for the whole job

**groundsella, grondsella, grunsella**, groundsill, foundation beam
**groundsellatio**, laying of groundsill
**groundsello**, to lay a groundsill
**grova** *see* **grava**
**grovetta**, small grove
**grudum** *see* **grutum**
**gruellum**, meal, esp. oatmeal
**grunsella** *see* **groundsella**
**grutum, grudum**, coarse meal for brewing
GRYPE *see* **gropa**
GRYPNAIL *see* GROPENAILE
**gula**, throat; **g. Augusti**, GULAUST, first day of August, Lammas
**gulfus, golfa**, pit
**gumphus, gunphus**, iron peg; part of door fastening
**gurges, gorgius**, whirlpool; weir
**guttera, goterum, gutterum**, drainage channel, gutter
**gyro, giro**, triangular piece of land; *cf.* **gora**
GYVES, shackles

# H

HACCHE, gate; wicket
**hachia**, HACHAT, hatchet, axe
**haedus, edis, hedus**, kid
**haia 1, aya, haya, heia**, hedge; enclosure in forest; *cf.* HAYE; **hesa**
**haia 2, heya**, main beam of plough
**haiagium, hayagium**, wood suitable for fencing
**haicia** *see* **hesa**
**haiwardus, haywardus, heiwardus**, HEYWARD, hayward
HAKENEY, HAKENEYE, hackney (horse)
HALFENDEAL, HALFENDELL, half part
**halimotus**, HALYMOTE, HELYMOT, manor court, 'hall moot'
HALK, HAULKE, HOLKE, corner; (?) hollow of land
**halla**, hall
HALLHOUSE, principal dwelling of farm
HALM, haulm, stubble; *cf.* HAULMING
**halmo**, to gather stubble; *cf.* **calamo**
HALPAN, (?) reinforcement; supporting beam
HALYMOTE *see* **halimotus**
**hama**, HAME, part of horse-harness
**hamatio**, hooking; reaping
HAME *see* **hama**
**hamelettum, hamelettus**, HAMELETTE, hamlet
**hamillus**, small (fish) hook
**hamo**, to hook
**hamsocna**, HAMPSOKEN, HAMSOKNE, house-breaking
**hamstallarius**, tenant of homestead
**hamstallum**, HAMSTAL, HEMPSTALL, HOMESTAL, homestead
**hamus**, hook
**handhabbenda**, HONDHABBANDE, (thief) caught w. stolen chattel
**handlum, handillum**, handle (of plough)
HANGE *see* HENGE
HANT, feeding place of animals
**hapsa** *see* **haspa**
**haracium 1**, stud for horses
**haracium 2**, HARAS, mixed crop; (?) peas, oats and beans
**harepipa**, HAREPYPE, snare
HAREWYNGSELVER, HARWYNGSELVER, (?) payment in lieu of harrowing service; but *cf.* HERYNGSELVER
**harnesio**, to equip; to decorate
**harnesium, harnesia**, harness
**harpex** *see* **herpex**
HARRE, hinge of door or gate
**harundinetum**, reed-bed
**harundo**, reed
HARWYNGSELVER *see* HAREWYNGSELVER
HARYNGSELVER *see* HERYNGSELVER
**haspa, hapsa**, hasp for door or window; part of plough harness
**haspo**, to fasten
**hastella** *see* **astella**
**hastiberum**, HASTYBERE, early barley; *cf.* **bera 2**

# HAULKE

HAULKE see HALK
HAULMING, YELMING, bundling straw for thatch; cf. HALM
HAUNDIRE, (?) andiron
HAVEDSCOT see HEVEDSHOT
**haverus** see **averus**
HAWYNGBLOK, chopping-block
**haya** see **haia 1**
**hayagium** see **haiagium**
HAYBOTE, AYBOTE, HEDGBOOTE, HEYBOTE, right to take wood for fencing
HAYE 1, hedge; enclosure in forest; cf. **haia 1**
HAYE 2, rabbit-net
HAYMONGESTRE, hay-dealer
**haywardus** see **haiwardus**
**hebdomadalis**, weekly
**hebdomas, ebdomas, hebdomada**, week
**hecca, hechia, hekca**, hedge, fence
HECHEWEY, gateway
**hechia** see **hecca**
**hedus** see **haedus**
**hedera, edera**, ivy
HEDGBOOTE see HAYBOTE
HEDSELLE see HEVEDSELLE
HEFKERE, heifer
HEGRAVE, HEGREWE, hedgerow
**heia** see **haia 1**
**heiro**, heron
**heironcellus, herunsellus**, young heron
**heisiamentum** see **aisiamentum**
**heiwardus** see **haiwardus**
**hekca** see **hecca**
HELYMOT see **halimotus**
HEMPSTALL see **hamstallum**
HENDERBAY, rear bay; cf. **baia 1**
HENGE, HANGE, hinge
**hengellum**, HENGEL, hinge
HENGENAIL, hinge-nail
HENGENETT, hangnet
**hensiclius** see **ensiculus**
**herba**, small plant; grass
**herbagium**, right of pasture; grass; hay
**hercia**, harrow; **h. clausa**, h. with iron cap; **h. duplex**, double h.; **h. simplex**, single h.; cf. **herpex**
**herciator**, harrower
**herciatura**, harrowing (service)

**hercio**, to harrow
HERDEL see **hirdellum**
**herdewica**, HERDWIKE, dairy farm
**herietum, heriettum**, heriot, death duty, right to take best beast or chattel
HERINGGRESE, fish-oil
**herpex, harpex, herpica**, harrow; cf. **hercia**
**hertlatha**, HERTELATH, lath of (?) heart-wood
**herunsellus** see **heironcellus**
HERYNGSELVER, HARYNGSELVER, payment in lieu of herrings; but cf. HAREWYNGSELVER
**hesa, haicia, hesia, hesium, heskia, heysia**, hedge, fence; cf. **haia 1**
**hevedclutum**, HEVEDCLUTE, metal plate on plough head
HEVEDLOND, headland
HEVEDSELLE, HEDSELLE, canopy over mill hoist; (?) mainsail of windmill
HEVEDSHOT, HAVEDSCOT, HEVEDSOT, poll-tax
HEVEDTROW, head beam of windmill
HEVYDSTAL, headstall; halter
**heya** see **haia 2**
HEYBOTE see HAYEBOTE
HEYRECLOTH, hair-cloth
**heysia** see **hesa**
HEYWARD see **haiwardus**
HEYWARDSHEPE, office of hayward
**hibernagium, hivernagium, hyvernagium, ivernagium, yvernagium**, winter crop
**hida, hyda, yda**, hide, us. 120 acres; cf. **virgata**
**hidagium**, hidage, land tax
**hidarius**, tenant of hide-land
**hiemalis, yemalis**, of or for winter
**hiems**, winter
HILLER see †**illera**
**hinniculus, hynniculus**, young fawn
**hinnulus**, fawn
**hirdellum**, HERDEL, hurdle
**hivernagium** see **hibernagium**
**hobelarius**, light horseman, hobbler
HOBBY, OBBEY, small horse, pony
**hockdies**, HOKEDAI, hockday, second Tuesday after Easter

**hocus, hokum, hokus**, HOK, hook; hook-shaped piece of land
**hoga**, 'how', hill
**hogaster, hogattus, hogerellus, hogettus, hoggus**, hogget, pig or sheep, us. in second year
HOGSHORSCHIP, sheep, (?) hogget after shearing
HOK *see* **hocus**
HOKEDAI *see* **hockdies**
**hokum, hokus**, *see* **hocus**
HOKYNG, hooking
HOLILOFE, bread given for distribution in church; *cf.* CHIRCHELOFE
HOLKE *see* HALK
**holmus, holma, hulma**, 'holm', river meadow
HOLTEYLL, gutter tile
**holus, holerum, olus**, vegetable
**homagium**, homage, act of allegiance; body of tenants
HOMESTAL *see* **hamstallum**
**homo**, man; **h. liber**, freeman
HONDHABBANDE *see* **handhabbenda**
HOOPSHIDE, HOPSHIDE, wooden strip for barrels
HOORE, hoar; white
**hopa 1**, HOPE, hoppet, piece of enclosed (marsh) land
**hopa 2, hoppa**, HOPE, dry measure
**hopa 3**, hoop
**hopella**, small hoppet
HOPENET, hoop-net
**hopetta, hopettus, hopetus**, small hoppet
HOPETYMBER, wood for hoops
**hoppa** *see* **hopa 2**
HOPSHIDE *see* HOOPSHIDE
**hora**, hour; **nona h.**, noon
**hordeaceus, hordaiceus, hordycius**, of barley, esp. w. ref. to brewing
**hordeum, ordeum**, barley; **h. palmale**, spring-sown barley
**horngeldum**, HORNGILDE, cornage, tax on cattle
**horreum, horrea, horreus, orreum**, barn, granary; **h. decimale**, tithe barn

HORSECOMBE, currycomb
HORSECORN, mixed feed w. oats
HORTOUR *see* **hurtellum**
**hortus, ortus**, garden
HOSEMUD *see* **osmondum**
**hospitium**, inn; house; household
**hospito**, to lodge (person); to store (goods)
**hostelagium, ostilagium**, lodging; tenants' duty to provide lodging
**hostia, hostium**, *see* **ostium**
HOSTYLEMENTS *see* **ustilamenta**
HOUCE *see* **hucia**
HOUSEBOTE *see* **husbota**
HOUSTER, (?) to harbour
**hovellus**, HOVELL, shed; hut; hovel
**howa**, HOWE, hoe
HOWYS *see* **hucia**
HOYNET *see* **oinonetta**
**hucia**, HOUCE, HOWYS, saddle cloth
HULEVRE, HULVER, holly
**hulma** *see* **holmus**
HUNDLACCHE, (?) dog-flap in door
HUNDREDSCOT, customary payment to lord of the hundred
**hundredum, hundredus**, hundred, division of shire
HUNTHIELD *see* UNYELDE
**hurtardellus, hurticulus**, young ram
**hurtardus**, ram
**hurtellum, hurturium**, HORTOUR, HURTOUR, 'hurter', metal plate between axle and wheel
**hurticulus** *see* **hurtardellus**
**husbanderia, husbandria, husbonderia**, tillage, husbandry
**husbandus**, householder
**husbota**, HOUSEBOTE, right to take wood for house repairs
**hussus**, holly
**hutesium, hutesia**, hue and cry
**hutlagatus** *see* **utlagatus**
**hyda** *see* **hida**
**hynniculus** *see* **hinniculus**
**hyvernagium** *see* **hibernagium**

# I

**igneus** *see* **domus**
**ignis**, fire
†**illera**, HILLER, elder tree
**imbladio**, to sow w. corn
**imbladitura**, putting under corn
**imparcamentum**, enclosure
**imparco**, to enclose; to impound
**impejoramentum**, damage
IMPETONE, (?) enclosure for saplings
**impignoratio**, pledge; mortgage
**impignoro**, to pledge; to mortgage
**impinguatio**, fattening
**impinguo**, to fatten
**implementum** (live) stock
**impo**, to plant, to 'imp'; *cf.* YMPE
**inanulatus**, (pig) not ringed
**includo**, to enclose; **inclusa, inclusum**, enclosure
**incrasso**, to fatten (animal)
**incrementum**, increase; extension (of tenement); intake
**incultus**, uncultivated, fallow
**indentatus** *see* **billa**
**infranco, infranchio**, to free, to enfranchise
**ingenium**, trap, snare
**ingistiamentum** *see* **agistamentum**
**ingrangio**, to garner, to store in grange
**ingressus**, entry
**inhabilis**, unfit; unbroken
INHEWE, YNHUWE, indoor servant
**inhorreatio**, garnering
**inhorreo, inhorrio, inorreo**, to garner
**inlanda**, INLAND, demesne land
**innamia, innamium**, INNAM, INNOME, intake, enclosure; good(s) distrained
INNED, YNNED, enclosed; *cf.* NEWYNNED
INNOME *see* **innamia**
INSETHOUS, farm building
**insidiatrix**, (female) busybody
**insidio**, to waylay
**instauramentum**, livestock
**interrogo**, to ask; to look for; to find
**intrinsecus**, on the inside, within; domestic
**intro**, to enter; to break in
**invadiatio**, pledging
**ivernagium** *see* **hibernagium**

# J

**jaceo**, to lie; **j. ad firmam**, to be farmed; **j. ad warectum**, to lie fallow
**jacio**, to throw
**jactatio**, tossing (esp. grain)
**jactitor**, tosser
**jacto**, to toss; to throw; **j. vitulum**, to drop a calf
JAGG, small (cart) load
**jalnetum** *see* **jaunetum**
**jampnum** *see* **jaunum**
**jantaculum** *see* **jentaculum**
**janua**, door; gate
**jaunetum, jalnetum**, gorseland
**jaunum, jampnum**, gorse
**jenectum** *see* **genesta**
**jentaculum, gentaculum, jantaculum**, breakfast
**jento, gento**, to breakfast
**jercia** *see* **gercia**
**jerma** *see* **germen**
**jugum**, yoke
**jumenta, jumentum**, mare
**juncaria, junkeria**, rush-bed
**juncheium**, rush-bed
**junco**, to cut or strew w. rushes
**junctor**, joiner, carpenter
**junctura**, joining, fastening; plough-team; wife's jointure
**juncus**, rush
**jungatio**, yoking

**jungo**, to join together; to yoke (plough or animal); *see also* **animal**
**junkeria** *see* **juncaria**

**juvencus, juvenculus**, bullock; **juvenca, juvencula**, heifer
JYSTE *see* **gista**

# K

KACCHE, catch; door-fastening
**kaeta** *see* **escaeta**
**karatative** *see* **caritative**
**karectata** *see* **carrettata**
**kario** *see* **carrio**
KARLOKE *see* CARLOKKE
**karukarius** *see* **carrucarius**
**katallum** *see* **catallum**
KAWSE *see* **calceta 1**
**kaya, caium, keya**, CAYE, quay, wharf
**kayo**, to build a wharf
**kedellus** *see* **kidellus**
**kemelinus** *see* **cumelina**
KENDALE, green woollen cloth, kendal
KETIL, kettle
KEVECELLE, KEVERCELLE, KEVESEL, kibble, horse's bit

KEVERCHEFWASCHER, washerwoman
**keya** *see* **kaya**
KEYLES, ninepins, kayles
**kidellus, kedellus, kydalus**, KYDELL, kiddle, fish-trap
**kipclutum** *see* **chipclutum**
**kippella** *see* **clippella**
**kippum**, KIP, bundle of hides, kip; *cf.* KYPPECORD
KNOWLEDGE MONEY, payment by tenant to new lord; *cf.* **recognitio**; (?) SNOTTERINGSELVER
**kydalus**, KYDELL *see* **kidellus**
KYDELCOTE, hut for kiddle-watcher
KYNNGESWARD, 'ward-penny'; *cf.* WARDEPENI
KYPPECORD, cord for tying kip; *cf.* **kippum**

# L

**labor**, work
**laborarius, laborator**, workman, labourer
**laboro**, to work
**lac**, milk
**lachettum, lachetum**, tie-beam
**lacista** *see* **lycista**
**lactabilis**, milch (cow)
**lactagium**, dairy produce; payment for pasturage
**lactatio**, milking
**lactator**, milker
**lacticinium**, dairy produce
**lacto**, to suckle; **lactans**, suckling
**lada 1, lata**, LEETE, lade, leat, watercourse
**lada 2, loda**, load
**lagehundredum**, law hundred
**lagena**, gallon; peck

LAMBHERDE, shepherd in charge of lambs
LAMME, 'lam', fishing net
**lampas**, lamp, lantern
**lana**, wool; **l. ventris**, belly-fleece; *see also* **receptor**; BELLEFLES; WAMBELOKES
**lanceo**, to debouch, to abut on
**landa, launda**, LAUNDE, forest clearing; open grassland
**landgabulum**, LANDGAVEL, land-tax
**lania, leyna**, LEYNE, oyster bed
**lanutus** *see* **pellis**
**lappata** *see* **loppo**
**laqueus, laquius**, tie-beam; snare
**lardarium**, larder
**lardarius, lardinarius**, larderer
**largio**, to enlarge
LASPANI, (?) payment for pasturing on green; *cf.* LEPSELVER, LESESELVER

**lastum**, LAST, last, load (of herrings)
**lata 1** *see* **lada**
**lata 2, latea, latha**, LATTHE, lath
LATHENAYL, LATTENAYL, LATTHENAYL, nail for fixing laths
**lathomus** *see* **latomus**
**lato 1, latho, latto**, to cover w. laths
**lato 2**, LATONE, LATTENE, brass, latten
**latomus, lathomus**, stone-mason
LATONE *see* **lato 2**
**latrina**, privy
**latro**, thief
**latrunculus, latruncula**, thief
LATTENAYL *see* LATHENAYL
LATTENE *see* **lato 2**
LATTHE *see* **lata 2**
LATTHENAYL *see* LATHENAYL
**latto** *see* **lato 1**
**launda**, LAUNDE *see* **landa**
**lavacrum**, wash-basin; ewer
**lavatio**, washing; *cf.* **lotio**
**lavator**, washer (esp. of sheep)
**lavatorium**, wash-basin
**lavendria**, laundry; *see also* **domus**
**lavo**, to wash (esp. sheep)
**laweo**, to 'law' (mutilate) animals
**lectus**, bed; base-frame; chassis
**ledga** *see* **lega**
LEEFSELE, LEFSEL, shelter made of foliage
LEETE *see* **lada 1**
**lega, ledga, leggum**, LEGGE, cross-bar (of door); strip, band (of hide or metal)
**leirwita, leriwitta**, LETHERWYTA, LEYRWYTE, fine for incontinence, 'leirwite'
LEPHACCHE, (?) deer-leap
LEPHENNE, hen given as rent
**leporarius**, greyhound
LEPSELVER, (?) payment for driving cattle over green; *cf.* LASPANI; LESESELVER
**leriwitta** *see* **leirwita**
LESESELVER, (?) payment for pasturing cattle on green; *cf.* LASPANI; LEPSELVER
†**lesio**, to separate (seeds of weeds from wheat)
**leta**, leet, division of hundred; court leet; *see also* **certus**

LETHERWYTA *see* **leirwita**
**leuca**, league, measure of distance
**levator**, loader (esp. of hay)
**levo**, to lift, to raise; *see also* **cubo**
**leyna** *see* **lania**
**leynagium, linagium**, revenue from oyster-laying
LEYNE *see* **lania**
LEYRWYTE *see* **leirwita**
LHADSOULE, (?) stake supporting (cart) load
**libella**, half-pound weight
**liber 1**, book; charter
**liber 2**, free; *see also* **bancus; homo; warenna**
**liberatio**, livery, allowance of food or clothing
**libertas**, privileged area, liberty
**libra**, pound (money or weight)
**librata**, pound's worth (of land or rent)
**lichinus** *see* **lychnus**
**licista** *see* **lycista**
**ligamen**, binding; strap
**ligatio, ligatorium**, binding; strap
**ligneus**, wooden
**ligniculum, liniculum**, small piece of wood; twig
**lignum**, wood
**ligo 1**, to bind (esp. sheaves)
**ligo 2**, mattock
**limo**, shaft (of cart)
**linagium** *see* **leynagium**
**linarium**, flax-ground
**linca, lincum**, LYNCE, linch pin; *cf.* **linkum**
**lingua**, tongue; *see also* **malus**
**liniculum** *see* **ligniculum**
**linkum** link (of chain); *cf.* **linca**
**lintellus**, (linen) sheet
**lintheamen, linthiamen**, (linen) sheet
**linum**, flax; linen
**litera, literium**, litter, bedding (for animals)
**litigatrix**, scold, quarrelsome woman
**littera**, letter
**locatio**, hiring
**loccum, lockum**, lock, weir; door-lock
**locellus**, box; compartment
**lockum** *see* **loccum**

**loco**, to hire
**locus**, place, district
**loda** *see* **lada 2**
LODCARTE, farm cart
LODCUMB, coomb load, four bushels; *see also* **cumba**
**lodex**, blanket
LODNAIL, (?) heavy nail, load-pin
LODOTEN, payment in oats, (?) in lieu of loading service
**logea, logia, logium**, LOGGE, lodge, hut
**loggio, loggo**, to make logs; to repair woodwork
LOKERE, look-out; herdsman
LOLLES, darnel
**longo**, to lengthen
**longus**, long
LOOM, clay, mud; *see also* **lutum**
LOPPES AND CROPPES, tops and branches
**loppo**, (?) **lappo**, to lop (trees); **lappata**, loppings

**lotio**, washing; *cf.* **lavatio**
**lottum, lottus**, lot, share of tax; *cf.* **scotum**
**lucrabilis**, profitable
**lucrum**, gain
**ludus, lusus**, game; **l. globorum**, game of bowls
**lupus**, wolf; **l. aquaticus**, pike
**lusus** *see* **ludus**
**luteus, lutosus**, of clay, earthen; *see also* **murus**
**luto, luteo**, to daub; to plaster
**lutum**, clay, mud; *see also* LOOM
**lychnus, lichinus**, lamp-wick
**lycista, lacista, licista**, greyhound bitch
LYMEBRENNER, lime burner
LYMEKELNE, limekiln
LYNCE *see* **linca**
LYSTE, 'list', border, edge

# M

**macella**, slaughterhouse, shambles
**macerinus** *see* **mazerinus**
**mactatio**, slaughtering
**macto**, to slaughter
MADER, madder
**madidus**, wet; *see also* **precaria**
MADLARD *see* **mallardus**
MADRAM *see* MEDRAM
**maeremium, meremium**, timber; *cf.* **materies; matrinum**
**magnus**, large; *see also* **petra**
MAL DE LANGE *see* **malus**
MALELOND, MOLLOND, **terra de** MOLL, molland, land on which rent is paid; *see also* **malmannus**; MOLE
**maliolus** *see* **malleolus**
**mallardus, mathelardus**, MADLARD, mallard, wild drake
**malleolus, maliolus**, small hammer
**malleus**, hammer
**malmannus, meelmannus, molmannus, moolemannus**, molman, rent-paying tenant; *see also* MALELOND

MALREPE, (?) payment in lieu of reaping
MALTPENS, MALTPANES, MALTPENYS, MALTSELVER, payment in lieu of making malt
**malum**, apple
**malus**, bad; **malum linguae**, MAL DE LANGE, foot and mouth disease
**managium**, house, household
**mancornum, mancorna**, 'mongcorn', mixed corn; grain for human consumption
**manduco**, to eat, to feed (on)
**maneriolum**, small manor
**manerium**, manor
**mangerium, maniorium**, MANGOUR, manger
**manimola** *see* **manumola**
**maniorium** *see* **mangerium**
**maniplus, manupulus**, handful, measure (of corn)
**mansa, mansus**, measure of land
**mansio, mansura**, house; tenement; manor
**mantica**, saddle-bag

**manualis** *see* **mola; molendinum**
**manucaptio**, bail
**manumitto**, to free (bondsman)
**manumola, manimola**, hand-mill
**manupastus**, member of household
**manupulus** *see* **manipulus**
**manura**, (?) **menuria**, tillage
**manus**, hand
MAPELE, maple tree
**mappa**, cloth; napkin
**mara**, lake, mere
**marca**, mark (13s. 4d.)
**marcata**, mark's worth (of land)
**marcatum** *see* **mercatum**
**marchalsia** *see* **marescalcia**
**marchia, merca, merka**, boundary; borderland, march
**marella** *see* **marla**
**marescalcia, marchalsia**, farriery
**marescallo, mareschello**, to treat sick animals
**marescallus**, farrier; marshal
**marola** *see* **auca**
**mariscus**, marsh
**maritimus** *see* **carbo**
MARKINGFLES, MARKYNGFLES, (fleece as) shepherd's perquisite
MARKINGHOG, (pig as) swineherd's perquisite
MARKINGLAMBE, (?) (lamb as) shearer's perquisite
MARKINGLOT, shepherd's perquisite
MARKSOIL, boundary mark
MARKSYLVER, tallage
MARKYNGFLES *see* MARKINGFLES
MARKYNGYRON, MERKYNGEYRYN, branding-iron
**marla, marella**, marl, clay soil; *cf.* MERELLHEAPE
**marlatura**; marling
**marlera, marlerium**, marl-pit
**marlo**, to apply marl
**marola** *see* **auca**
**marra** *see* **murra**
**martellum**, hammer
**masagium** *see* **mesuagium**
**mastinus, †mautivus**, MESTYFF, mastiff
MASTYGREHOUNDE, (?) mastiff - greyhound

**materies**, (?) timber; *see also* **maeremium; matrinum**
**mathelardus** *see* **mallardus**
**matinellum, matutinellum**, lunch
**matrinum**, (?) timber; *cf.* **maeremium; materies**
**matrix** *see* **ovis**
**mattokus**, MATTOK, mattock
**matutinellum** *see* **matinellum**
**†mautivus** *see* **mastinus**
**mazerinus, macerinus**, of maple wood
MEAKE, MEYKE, long-handled hook for cutting peas
MEDEWEMETEBENE, mowing boon-work
**medicina**, medicine; medical treatment
**medietas**, moiety, half; middle part
**medius**, middle; intermediate
MEDRAM, MADRAM, meadow ram, perquisite of mower; *cf.* RAMSELVER
MEDWERITH, MEDWESILVER, payment in lieu of mower's perquisite
**meelmannus** *see* **malmannus**
**meia**, 'mow', stack
**meisa**, MEYS, 'mease', measure of herrings
**mel**, honey
MELLECROW *see* MELNECROW
MELLEDAM, mill-dam
MELNECOT, MILNECOTE, mill-house
MELNECROW, MELLECROW, pick for dressing mill-stone
MELTING, MELTYNG, malting
MENEGOS, goose given for harvest supper; *cf.* REPGOS
MENESCHEF, sheaf given (?) for carrying service; *cf.* BENSCHEF
**mensa**, board; table; food
**mensalis** *see* **tabula**
MENSION, MENTION, trace, indication
**mensura**, measure; **m. cumulata**, heaped measure; **m. rasa**, razed measure
MENTION *see* MENSION
**menuria** *see* **manura**
**merca** *see* **marchia**
**mercatorius**, of a market; mercantile
**mercatum, marcatum**, market
**merchetum**, MERCHAT, 'merchet', payment to lord on marriage of tenant's daughter

MERELLHEAPE, marl-heap; *cf.* **marla**
**meremium** *see* **maeremium**
MERESTAKE, boundary mark
**merica** *see* **myrica**
**merka** *see* **marchia**
MERKYNGEYREN *see* MARKYNGYRON
**mescinga**, food rent; *cf.* MESYNGSELVER
**meskenninga** *see* **miskenninga**
MESSAGER, boat
**messio**, reaping
**messor**, reaper
MESSYNGKNIFE, table knife
MESTYFF *see* **mastinus**
**mesuagium, masagium, messuagium**, messuage, house
MESYNGSELVER, (?) a food rent; *cf.* **mescinga**
**meta**, METE, boundary
**metatus** *see* **metor**
METE *see* **meta**
METECORN, grain allowance
METELES, without food (of days worked)
**meto**, to reap
**metor**, to measure; **metatus**, measured
METYERD, measuring rod
MEYKE *see* MEAKE
**meylo** *see* **mullo**
MEYS *see* **meisa**
**mica, michia**, MICHE, small loaf
**mille**, a thousand; **milia (passuum)**, a mile
MILNECOTE *see* MELNECOT
**mina**, dry measure
**mino**, to drive (cattle)
**minutus**, lesser (tenant); cut or chopped; *see also* **sal**
**miskenninga, meskenninga**, 'miskenning', mistake in pleading
**mixtilio, mixtura**, mixed corn
**mobilis** *see* **redditus**
**modellus, modiolus** *see* **modulus**
**modius**, dry measure (? peck); liquid measure (? 8 gallons)
**modulus, modellus, modiolus, muellus**, nave (of wheel)
**mola**, mill; millstone; **m. manualis**, hand-mill
**molarius**, for grinding
**molo**, to grind

MOLDYNGBORD, (tiler's) moulding-board
MOLE, MOLEACRE, service or rent; *cf.* MALELOND
**molendinarius**, miller
**molendinum**, mill; **m. aquaticum**, water-mill; **m. caballarium, equitium, equorum**, horse-mill; **m. folerectum, fullonicum**, fulling-mill; **m. manuale**, hand-mill; **m. ventricium, ad ventum**; windmill; *see also* **posta**
MOLL, MOLLOND *see* MALELOND
**mollis**, soft; *see also* **bladum**
**molmannus** *see* **malmannus**
**molossus**, hound, 'ban-dog'
**moncellus, muncellus**, small pile
**mondinum**, 'Monday land'; *cf.* MUNDAYLOND
MONGWODE, (?) mixed woodland
**mons**, hill; **m. talpae**, molehill
**montanus**, hilly
†**montura** *see* **mortuarium**
**moolemannus** *see* **malmannus**
**mora 1**, marsh; moor
**mora 2**, delay; residence; *see also* **traho**
**morbosium**, dead wood
**morina**, murrain
**morior**, to die; **mortuus**, dead; *see also* **sepes; staurum**
**mors**, death
**mortarium, morterium**, mortar
**morterellus, mortrellus**, pudding, mortrewe
**morterium** *see* **mortarium**
**mortrellus** *see* **morterellus**
**mortuarium**, †**montura**, mortuary payment (often an animal)
**mortuus** *see* **morior**
**mota**, moat
MOWYNGSELVER, (?) payment in lieu of mowing service
**mucaria**, (?) peat-moss; *cf.* **muscus**
**muellus** *see* **modulus**
**mulcibilis**, (?) in milk (of cow)
**mulcum** *see* **mulsum**
**mullo, meylo, mullio**, hay-cock
**mulsum, mulcum**, honey drink; (?) mead
**multo**, wether, sheep

**multura**, multure, toll for corn-grinding
**mulus**, mule
**muncellus** *see* **moncellus**
MUNDAYLOND, land for which tenant did service on Mondays, 'Monday land'; *cf.* **mondinum**
**mundo**, to clean
**murex**, shellfish
**murra, marra**, mazer bowl
**murus**, wall; **m. luteus**, of clay
**muscarium**, fly-whisk
**musculus**, mussel
**muscus**, moss; *cf.* **mucaria**
MUSTREDEVELERES, MUSTERCLOTH, musterdevillers, grey woollen cloth
**myrica, merica**, shrub, (?) broom

# N

**namium**, (goods seized in) distraint
**nativus**, villein, bondsman; *see also* **terra**
**natta, natus**, (reed) mat
NAVE, hub (of wheel)
**navis**, ship
NEDELE, needle; wooden prop (in mill); (*vb.*) to provide props
**nemus**, grove
NEWYNNED, newly enclosed; *cf.* INNED
**nobile**, noble (gold coin)
**noctivagus**, night prowler
**nodus**, knot; knob
**nonus**, ninth; *see also* **hora**
**norma**, rule, regulation
NORTHENSTERES, cattle from the north
**nostla**, NOSTELE, 'nostel', nose-band of harness
NOTE, (metal) nut
**nudus**, bare; *see also* **carretta**
**numero**, to count; *see also* **pecunia**
**nummus**, coin, penny
**nundinae** (*pl.*), fair
**nutrimentum**, fodder; rearing
**nutritura**, fodder; rearing
**nux**, nut; **n. Walliae**, walnut (*i.e.* Welsh or foreign nut)

# O

OBBEY *see* HOBBY
**objurgatrix**, scold, shrew
**oblatio**, offering, payment
**oblongus**, (?) elongated
**obolata, oblata**, (land yielding) a halfpenny rent; halfpennyworth
**obolus**, halfpenny
**obstupo**, to block
**obturo, opturo**, to stop up
**oculus**, eye; **o. petrae**, hole in centre of millstone
**odorisequus, odorencicus, odorinsequus**, tracker dog
OFFRYNGPANY, OFFRYNGSELVER, penny given to lord at Christmas; *cf.* CRISTEMASSECUSTOM
**oinonetta**, HOYNET, onion, shallot
OLBE *see* OLVE
**olla**, pot, measure, **ollae custumariae**, custom-pots, ale-toll; *cf.* CUSTOMPOTTES
**ollula**, small pot
**olus** *see* **holus**
OLVE, OLBE, part of mill-wheel; *cf.* **whola**
OMBLYNGSHON (*pl.*) horseshoes (for 'ambler')
**oneratio**, loading
ONGIELL *see* UNYELDE
**onus**, load
OPELAUNDE, upland
**opella**, workshop

**operabilis**, workable
**operarius**, worker
**opero**, to work
**opturo** *see* **obturo**
**opus** (*pl.* **opera**), work, esp. labour service
**ordeum** *see* **hordeum**
**orreum** *see* **horreum**
**ortus** *see* **hortus**
OSEMUND *see* **osmondum**
**osera**, osier, willow
**osmondum**, HOSEMUD, OSEMUND, Swedish iron, osmund
**ostiarius**, door-keeper
**ostilagium** *see* **hostelagium**
**ostilmentum** *see* **ustilamenta**
**ostium, hostia, ostia**, door; *see also* **folium**

**ostrea**, oyster
**osturcus** *see* **asturcus**
OTEMELE, oatmeal
OTER, otter
**otiosus**, not working
OVELOTIREN, (?) pan for cooking eggs
OVERMAN, foreman
OVERWEY, beam over door
**ovile, ovilium**, sheepfold
**ovinus**, of sheep
**ovis**, sheep; **o. matrix**, ewe
**ovum**, egg
OWELE, awl
OXGANG, bovate
OYSERHOPE, osier-bed

# P

**paccator, pakkator**, (wool) packer
**paco**, to pay
PACTHRED, PAKTHERD, pack-thread
**padnagium** *see* **pannagium**
**pagius, paiettus**, page, servant
**pakkator** *see* **paccator**
PAKKESADILL, pack-saddle
PAKTHERD *see* PACTHRED
**palasium, palicium**, PALYS, paling; fence
**palea, pallea, pallium**, chaff
**palefridus, palfridus**, palfrey
PALFREYSILVUR, money for palfrey (as gift to new lord)
**paliciator**, park-keeper, paliser
**palicium** *see* **palasium**
**pallea, pallium**, *see* **palea**
**palma 1**, palm (of hand)
**palma 2**, 'palm', willow branch; *see also* **dies; ramus**
**palmalis** *see* **hordeum**
**palmo**, to grasp
**palus 1**, paling, fence
**palus 2**, marsh
PALYS *see* **palasium**
**pandoxator**, brewer
**pandoxatrix**, ale-wife
**panellus**, saddle-pad

**panis**, bread; **p. benedictus**, bread blessed for distribution; **p. equinus**, horse-bread; **p. uncti**, loaf of lard
**pannagio**, to pay pannage
**pannagium, padnagium, pasnagium, pounagium, pownagium**, pannage, payment for pasturing pigs
**pannarius**, draper
**panniculus**, small cloth
**pannus 1**, cloth
**pannus 2**, squared timber
PAP, custard, soft pudding
**papyrus, paupirus**, paper (book)
**par**, a pair; **p. compedum**, pair of fetters; **p. stipitum**, pair of stocks
**Parasceves** *see* **dies**
**parca** *see* **parcus**
**parcarius**, parker
**parcella**, piece of land
**parcenarius**, joint tenant, parcener
**parchetto** *see* **pargetto**
**parco**, to empark
**parcus, parca**, park, enclosure
**pardix** *see* **perdix**
PARDON SONDAY, Sunday before Lent
**pargetto, parchetto, parjetto**, to parget, to plaster

**parlorium**, parlour
**parna** *see* **perna**
**paro**, to prepare; to trim
**paropsis, perapsis**, dish
**parrocus**, PARROK, paddock
**partica** *see* **pertica**
**particula**, small piece (of land)
**parura**, trimming
**parvus**, small; *see also* **petra**
**Pascha**, Easter; **P. floridum**, Palm Sunday; *see also* **dies**
**pasco**, to feed; to graze
**pascua**, pasture, 'feeding'
PASKES, bearings (under head of mill axle)
**pasnagium** *see* **pannagium**
**passagium**, ferry; ferry toll
**passus**, pace, step; *see also* **mille**; **p.equorum**, horse-track at horse-mill; 'gang' (of mill spindles)
**pastor**, shepherd
**pastoragium**, pasture; *see also* **communitas**
**pastura**, (right to) pasture; *see also* **communa**
**pastus**, fodder; meal
**Paulinus** *see* **pollex**
**pauper**, pauper
**paupirus** *see* **papyrus**
**pava**, peahen
**paviculus, †faviculus**, young peacock
**pavo, phavo**, peacock
**pax**, peace; *see also* **perturbator**
**pecia**, piece
**pecora** *see* **pecus**
**pecunia**, money; **p. numerata**, cash; **p. viva**, livestock
**pecus** (or *pl*. **pecora**), livestock, cattle
**pedalis**, (for) foot; *see also* **ferrum; tabula; via**
**pedestris** *see* **averagium; via**
**pedito**, to tread, to walk
**pedo**, to fit (plough) w. foot
PEESUN *see* **pesso**
PEKE *see* **pica**
**pelleta, pelletta**, skin, pelt
**pellicium**, leather garment
**pelliculus**, small pelt
**pelliparo**, to dress skins
**pelliparus**, skinner
**pellis**, (animal) skin, pelt; **p. lanuta**, sheepskin
**pelvis**, basin, dish
**penarium**, cupboard; storehouse
**pendens, pendibilis** *see* **sera**
**pendo**, to hang; *see also* **sera**
**penetrale**, gimlet
**penetro**, to pierce
**penna**, feather; quill
**pennatus**, fledged (of arrow)
**pensa**, unit of weight, wey
**penticium**, lean-to; shed
**penula**, hood; fur edging
PEPYRQUERNE, pepper-mill
PERAMBUL. PERAMBOL, passage-way
**perambulo**, to walk around; to beat the bounds
**perapsis** *see* **paropsis**
**perautumpno**, to perform or supervise harvest work
**perca**, perch (measurement); *cf.* **pertica**
**perdix, pardix**, partridge
**pergamenum**, parchment
**perhendino**, to sojourn; to tend
**perna, parna (baconis)**, ham
**perpetro**, to acquire
**perquisitor, perquisitrix**, purchaser
**pertica, partica**, perch (length); *cf.* **perca**
**pertinentia**, appurtenances
**perturbator, perturbatrix (pacis)**, disturber (of the peace)
**pes**, foot, standard measure; base; socket of mill-spindle; *see also* POWLESFEET
**pescagium** *see* **piscagium**
PESEMONG, mixed crop of peas and beans; *cf.* BENEMONG
PESEREEK, pea-stack
**pesso, pessona, pessonium**, PEESUN, PESSUN, acorns, mast (food for pigs); pannage
**pestilencia**, the Plague; *see also* **tempus**
**peta**, peat
**peteus** *see* **puteus**
**petilium**, bird-bolt (arrow)
PETINGE, PETYNG, PETYNGGE, peat
**petra**, stone; stone weight (us. 14 lb.); **p. magna**, (?) full stone; millstone; *see also* **oculus**; **p. parva**, half-stone; **p. rubra, rubea**, red ochre

PETYNG, PETYNGGE *see* PETINGE
**phalera** (*pl.*), horse-trappings; *cf.*
   **fallaratus**
**phasianus, fesianus**, pheasant
**phavo** *see* **pavo**
**philum** *see* **filum**
**pica**, PEKE, pike (fish)
**picagium, pykagium**, 'pickage', payment for pitching market stall
**picchator**, one who pitches
**piccho, piko**, to pitch (hay)
PICHERE, PICHCHERE, one who pitches
**picosia,** †**picus**, PICOYSE, pickaxe
**pictillus** *see* **pitellum**
†**picus** *see* **picosia**
**pightellus, pihtlus** *see* **pitellum**
PIK, spikenail
**pikerellus**, small pike (fish)
PIKNAYL, PYKNAYL, spikenail
**piko** *see* **piccho**
**pila 1**, channel in tidal river, 'pill'
**pila 2**, ball
**pilarium** *see* **pillorium**
**pilium** *see* **pilleus; pilum**
**pilleus, pilium**, cap
PILLEWERE, pillowcase
**pillorium, pilarium**, pillory
**pilum, pilium**, bolt for crossbow
**pilus**, hair; horsehair
PINFALD *see* PUNDFALD
**pinguedo, pingwedo, pynguedo**, grease; fattening; season of grease (for venison)
**pingueo**, to fatten
**pinna**, pin; peg
**pinno**, to pin; to peg
**pipa**, pipe; cask
PIPES (*pl.*) piping (of harness)
**piper**, pepper
**pirreta**, perry
**pirum**, pear
**pirus**, pear-tree
**pisa**, pea
**piscagium, pescagium, piscaria**, fishery; fishing rights
**piscenarius**, fishmonger
**piscis**, fish
**pistor**, baker
**pistorium, pistrinum**, bakery; bakehouse

**pistrix, pixtrix** (*f.*), baker
**pitellum, pictillus, pightellus, pihtlus, pitellus**, pightle, small field
**pix**, tar, pitch
**pixis** *see* **pyxis**
**pixtrix** *see* **pistrix**
**placea**, plot (of land); site
**placitum**, plea; *see also* **domus**
**planca**, PLANKE, plank, board
PLANCHE, (plate for) horseshoe
**planco**, to plank; to floor
**planistra, planities**, plain; open ground; common; *cf.* **planum**
**planta**, plant; seedling
**planto**, to plant
**plantula**, shoot, young plant
**planum**, plain; open ground; *cf.* **planistra**
PLASH 1, PLASHE, PLASSHE, ford; watersplash
PLASH 2, to interweave, to pleach (hedge)
**plastro**, to plaster
**platea**, plate
PLAUNCHNAIL, plank-nail
**plaustratum**, waggon-load
**plaustrum**, waggon
PLAYSTALL, PLEYSTOW, playground
**plegium**, pledge; bail
**plegius**, pledger; **p. capitalis**, headborough
PLESAUNCE, special royal subsidy
PLEYSTOW *see* PLAYSTALL
**plica, plita**, PLYTE, plait; fold
**plico**, to bend; to twist; to fold
PLOCLOUT, iron plate at side of plough
PLONKETT, plunket, blue or grey woollen cloth
**plumbum**, lead; leaden vessel
PLYTE *see* **plica**
**poca**, POKE, bag, pouch
**pocenettus, pocinetum**, POCENET, POSCINET, POSSENETT, posnet, metal pot
POKE *see* **poca**
**polecta, polectus**, *see* **pulleta**
**politridium, politidium**, sieve; bolting-cloth
POLLARD 1, bran

POLLARD 2, POLLENGER, pollard tree
**pollex**, thumb; inch; **p. Paulinus**, inch by standard of St. Paul's, London; *cf.* POWLESFEET
**pollexa**, poleaxe
POLMARKED, marked on the head
POLYVE, (?) pulley
**pomarium, pomerium**, orchard
**pomum**, apple
POND, enclosure; (?) pasture; *cf.* **punda**; PUNDFALD
**ponda** *see* **pondus; punda**
**pondagium, ponnagium, pundagium**, poundage; toll
**ponderator**, inspector of weights
**pondus, ponda**, weight; load, 'wey'
**pons**, bridge
POPELER 1, poplar tree
POPELER 2, spoonbill
**porcaria, porcheria, porkeria**, pigsty
**porcarius**, swineherd
**porcello**, to farrow
**porcellus, porculus, purcellus**, piglet
**porcheria** *see* **porcaria**
**porcus**, pig; boar; **p. aetatis**, full grown pig; *see also* **cauda; comestus; domus**
**porectum** *see* **porretum**
**porkeria** *see* **porcaria**
PORPASSE, PURPOYS, porpoise
**porretum, porectum**, leek
**porta**, gate; **p. flottarum, fluvialis**, floodgate; *cf.* FLOTGATE
**portator (bursae)**, fair-time official
PORTMANMOTE, borough court
PORTSTRATE, PORTWEY, portway, highway
POSCINET, POSSENETT *see* **pocenettus**
**posta, postis**, post (of mill); **p. principalis**, king-post
**potagium 1**, drink money
**potagium 2**, pottage, broth
**potellus**, pottle, liquid measure
**pottarius**, potter
**pounagium** *see* **pannagium**
POUNCHEN, PUNCHEN, puncheon, liquid measure
POUNTFOLDE *see* PUNFALD

POWEL, pole
POWLESFEET, feet by standard measurement of St. Paul's, London; *cf.* **pes; pollex**
**pownagium** *see* **pannagium**
POY, gallery outside house
†POYNE *see* **pugnata**
**praecepes** *see* **praesepium**
**praedium, predium** estate; property
**praepositus, prepositus**, reeve
**praesepium, praecepes, praesepia, presepium**, stall (for cattle); manger
**praestatio, prestatio**, payment
**pratellum**, small meadow
**pratum**, meadow; **p. salsum**, salt-meadow
**pre-** *see also* **prae-**
**precaria, precarium, precatio**, boonwork; **p. amoris**, love-boon; **p. madida**, boonwork with ale; **p. sicca**, boonwork without ale
**pressorium, pressura**, press for cheese, cider or wine
PRIME, to prune (tree)
**principalis**, main, principal; *see also* **posta**
**priso, prisona**, prisoner
**privata** *see* **camera**
**privilegium**, (area of) jurisdiction
**profundus**, deep
**propars, purpars**, purparty, share
**prosterno**, to fell
**prostro**, to fell
**provendarium, provenderium**, food; fodder; measure of corn
**provisor**, surveyor; steward; overseer
**pugnata, pugillata**, †POYNE, fistful
**pugnus**, fist
**pulcinus**, chicken, pullet
**pulla, pullus**, chicken
**pullagium**, (?) fowl-rent
**pullanus**, foal, colt
**pulleta, polecta, polectus**, chicken, pullet
**pulmentum**, pottage
**pulteria**, poultry
**pulterius**, poulterer
PUNCHE, draught-horse, punch
PUNCHEN *see* POUNCHEN

**punctum**, point, tip
**punda, ponda**, pound, pinfold; *cf.*
   POND; PUNDFALD
**pundagium 1**, fine for impounded animals
**pundagium 2** *see* **pondagium**
PUNDES, rent; (?) pound-rent
PUNDFALD, PINFALD, POUNTFOLDE, PUNTFOLD, pound, pinfold; *cf.* POND; **punda**
**purallium**, purlieu
**purcellus** *see* **porcellus**
**purpars** *see* **propars**
PURPOYS *see* PORPASSE

**purprestura**, purpresture, encroachment
**puteus, peteus, putea**, pit; well
**putredo**, rot; rotten matter
**putrificio**, to rot
**putura 1**, food allowance
**putura 2**, (?) PUTYER, beam
PYBOT, (?) measure
PYHTWERREN, (?) pitch
**pykagium** *see* **picagium**
PYKNAYL *see* PIKNAYL
PYLEWHEY, (?) whey produced in cheese-making
**pynguedo** *see* **pinguedo**
**pyxis, pixis**, box, casket

# Q

**quadragesima**, Lent
**quadragesimalis**, Lenten
**quadrans**, farthing
**quadriga**, wagon
**quarentena**, furlong; *cf.* **ferlingata**
QUARTERBORDE, (?) wood for panelling
**quarterium, quartrona, quartronus**, QUARTROUN, QUARTROWN, quarter, fourth part; measure (weight or number)

**querculus**, oak sapling
**quercus**, oak
**quietus**, quit, balanced (of accounts)
QUILLET, small piece of land
**quinarius**, containing five
**quindecima**, (tax of) a fifteenth
**quisquiliae** (*pl.*), chips, loppings
**qwarva** *see* **wharva**

# R

**raccus** *see* **racka**
**racemus**, cluster; **r. zinziberis**, 'race', root of ginger
**racka, raccus**, feeding rack
**radius**, spoke; furrow
**rado**, to raze (measure of grain); *cf.* **mensura**
RAKE, rough or narrow path
†**rakio**, (?) to rake
**ramalia, ramalla, ramilla**; lopwood
**rammatio**, ramming
**rammo**, to ram; to pack
†**rammulus**, young ram
†**rammus**, ram

RAMSELVER, payment to mowers in lieu of customary ram; *cf.* MEDRAM
RAMSONE *see* RASEN
**ramus**, branch; **rami Palmarum**, Palm Sunday
**ranga, renga**, row, strip
**rasarium**, razed measure of grain
RASEN, RAMSONE, RASNE, RASONE, RESENE, wall plate
**rasorium**, razor
**rastellus**, small rake
**rastratio**, raking
**rastro**, to rake
**rastrum**, rake; hoe

RAWBAKKIS, type of (?) unfulled russet cloth
RAYLE, rack; rail
RAYLEBASE, rail-base (of framed building)
**rebinatio, rubinatio**, reploughing
**rebino, ribino, rubino**, to replough
**receptamentum, recettamentum**, harbouring (criminal)
**recepto, recetto**, to harbour
**receptor**, receiver; **r. lanae**, (official) receiver of wool
**rechacio**, to drive back (cattle)
RECHTREE, ridge-tree, horizontal timber on roof
**reclavo**, to renail
**recognitio**, recognition, acknowledgement of lordship; payment; *cf.* KNOWLEDGE MONEY; SNOTTERINGSELVER
**recordum, recurdum**, record; entry on roll
**recubo**, to re-lay
**recurdum** *see* **recordum**
**redditus**, rent; **r. assisus**, rent of assize, fixed rent; **r. mobilis**, rent in kind; **r. resolutus**, rent paid out by landlord
REDECUL, (? inflammatory) disease of swine
REDEFLEKKED, roan
REDELE, riddle, sieve
REDEN, RIDDEN, RYDEN, land cleared from waste
**redigo**, to bring back
REDINGE, REDYNG, red ochre, ruddle (for marking sheep)
**refocillamen, refocillatio**, refreshment
**refundo**, to refund
**refusio**, refunding
**regardator**, regarder, forest official
**regardum**, reward; payment
REGGEROPE, REGROPE, ridge-rope, saddle-band
REGGETYE, RIGGETEYE, ridge-rope
**regius**, royal; *see also* **via**
**regraciator, regraciatrix**, regrator, retailer
REGROPE *see* REGGEROPE

**reina, rena**, REYNE, rein
REKHAE, REKHAHE, rick-yard
**relevium**, feudal relief, payment on succession to holding
**relevo, relevio**, to redeem by relief
**relucrum**, aftermath, second crop (of hay); *cf.* **rewaynum**
**remaneo**, to remain; **remanens**, remainder (in account)
REMBLE, bundle (of hemp)
**rena** *see* **reina**
**renga** *see* **ranga**
**rentale**, rent book
**rentaria, renteria**, rented holding
**rentarius**, rent collector; rent payer
**reono**, to furrow; to dig channels
**reparo**, to repair
**repastus**, meal
**reperevus**, REPEREVE, REPEREWE, REPREVE, RIPEREVE, reap-reeve
REPGOS, REPGOOS, REPPGOS, RIPGOS, harvest supper; *cf.* MENEGOS
**reprisa**, deduction
REREBAY, back bay; *cf.* **baia 1**
**rescussus**, 'rescue' (esp. of cattle in distraint)
RESENE *see* RASEN
**resolutus** *see* **redditus**
RESSET *see* **rosetum**
RESSHESELVER, RUSCHEWSYLVER, payment in lieu of rushes for floor
**restauramentum**, re-stocking
RESTCLUT, RIESTCLUT, RISTCLUT, plate under plough-reest
RESTESHO, RESTSCHO, base of plough-reest
**resumptio**, taking back
**rete, rethium**, net
RETHERPENY, herbage rent, ox-penny
**rethium** *see* **rete**
**reticulum**, small net
**reto**, to 'ret', to soak (flax)
**retractio**, withdrawal
**retraho**, to withdraw; to withhold; to draw back
**retroherbagium**, second growth of grass
**rewaynum, rowannum, ruannum**, REWAIN, ROWAYN, second crop of hay; *cf.* **relucrum**

REY, 'race', measure (of garlic)
REYGNE see **rina**
REYNE see **reina**
**ribba, rybba**, bar (of gate); hurdle
**ribino** see **rebino**
RIDDEN see REDEN
RIESTCLUT see RESTCLUT
RIGG, (?) to fix in place
RIGGETEYE see REGGETYE
**rina**, REYGNE, RYNE, 'rind', iron support of upper millstone
**ringo**, to ring; to hoop
**ripa**, bank; wharf
**riparia, riperia**, river
RIPEREVE see **reperevus**
RIPGOS see REPGOS
†**ripplio**, RIPPLE, to comb-out seeds from flax
RIPREVE see **reperevus**
RISSET see **rosetum**
RISTCLUT see RESTCLUT
**rixatrix**, scold
**rochia, rochea**, roach (fish)
**roda**, rood, measurement of land
RODFALL, woodland cut at different time from the rest
ROFTEYL, roof (? ridge) tile
ROKHERDE, rook-scarer
ROMPANI, ROMSCOT, Peter's pence
RONGE see **runga**
**ropa**, ROPE, rope; ROPE MONDAY, Hock Monday, second M. after Easter
**rosa**, rose
**roscus** see **rusca 2**
ROSEL, ROSILE, ROSYLE, rosin
**rosetum**, RESSET, RISSET, bed of rushes

ROSILE, ROSYLE see ROSEL
**rota**, wheel
**rotabilis, rotatus**, wheeled; see also **civera**
**roto**, to revolve; (?) to fit w. wheel
**rotulus**, roll; see also **combustio**
**rowannum**, ROWAYN see **rewaynum**
ROWELE see **ruellus**
**ruannum** see **rewaynum**
**rubedo**, reddening
**ruber, rubeus**, red; see also **petra**
**rubigo**, rust
**rubinatio** see **rebinatio**
**rubino** see **rebino**
**rubus, rubius**, thorn, bramble
**rudus**, †**ruderia**, rotten material; rubble; ruins
**ruellus**, ROWELE, cogged wheel of mill
**rumor** see **tempus**
**runca**, brier
**runcinus**, rouncey, nag
RUNDLET, small cask
**runga**, RONGE, rung; stave
**rungatio**, fitting w. rungs
**rungo**, to fit with rungs
**rusca 1, ruscha**, beehive
**rusca 2, roscus**, rush (plant), rushes (coll.)
RUSCHEWYSYLVER see RESSHESELVER
**rusticus**, countryman, rustic
**rybba** see **ribba**
RYDE, 'ride', strap of hinge
RYDEN see REDEN
RYNE see **rina**
RYP, 'rip', wicker basket
RYVE, 'rive', rake

# S

**sabulum, zabulum**, sand; gravel
**sacculus**, small sack
**saccus, sacca**, sack; dry measure
SADECOTE, seed-lip, seed-basket
**sagena**, seine, fishing net
**sagimen**, fat, grease
**sagina**, fat

**sagium, sageum, sagum**, bolting-cloth (for refining flour)
**sagitta**, arrow
SAILCLOTHES, canvas for windmill sails
**sais-** see **seis-**
**saisona** see **seiso**
**sal**, salt; **s. minutus**, fine salt

# SALICETUM

**salicetum, saucetum, sauctetum**, willow-bed
**salicia** *see* **salix**
**salignus**, made of willow
**salina**, salt-pan
**salix, salicia**, willow
**salopus, sedlopum**, SEADLIP, seedlip, shallop
**salsa**, salt water; salt pan
**salsamentum**, sauce
**salsarium, sausarium**, salt-cellar
**salsugineus**, salty; briny
**salsum** *see* **fenum; pratum**
SALTCOT, SALTHOL, salting; salt-house
SALTFAT, SALTYNGFAT, brine tub
**salvacustodio**, to keep safe
SALVAGE, selvage, marginal land
**salvagius**, wild
**salvus**, safe
SAPLATHE, lathe made of sapwood
**sapo, savo**, soap
**sapula** *see* **scapula**
**saracra** *see* **seracra**
**sarateca** *see* **chirotheca**
**sarculatio, sarculatura**, hoeing
**sarculo, carculo, sarclo, serculo**, to hoe; to weed
SARGE, serge
SARPLARE, SARPULARE, wrapper for wool; bale
**sarra** *see* **serra**
**sarta** *see* **essarta**
**sartrinium, sartrinum**, tailor's shop
**satio**, sowing, planting
**saucetum, sauctetum** *see* **salicetum**
SAUDERE, solder
**sausarium** *see* **salsarium**
**sauvagina, savagina**, beasts of the chase
SAVE *see* SHAVE
**savo** *see* **sapo**
SAWYNGPET, sawing pit
**scabea, scabia**, scab (of animals)
**scabellum**, stool
**scabiosus, scabriosus**, scabby
**scala, schala**, ladder; **s. ad carrettam**, extension to cart
**scalarium, schalera**, stile
**scalonia**, scallion
**scamnum, scannum**, bench; bank
**scancile**, stile

**scannum** *see* **scamnum**
**scapha**, bowl
**scapula, sapula**, trimmings (of timber)
**scapulo, scapilo, scarpulo**, to trim, to dress (timber); *cf.* **sculpo**
**scella** *see* **sella**
**sceptum** *see* **septum**
SCHAKCLOT, sackcloth
**schala** *see* **scala**
**schaldaria** *see* **escaldaria**
**schalera** *see* **scalarium**
**scharum** *see* **sharrum**
SCHELDERESTE, SCHELDRESTCLUT, (?) metal sheathing for mould board of plough
SCHEPETORN, service connected w. lord's sheep
SCHEPSELVER *see* SHEPSILVER
SCHOFNET, SCOFNETT, shove-net, broad-mouthed net for river fishing
SCHOKK *see* **shockum**
SCHON (*pl.*), shoes
SCHOOT, SCOTE, shot, strip (of land)
SCHORE, sea-wall
SCHOTBORD, shotboard, planking
†**schrido** *see* **shreddo**
SCHYNGEL *see* **scindula**
**scicera** *see* **sicera**
**scindo**, to cleave; to split
**scindula, cendula, cindula, sindula**, SCHYNGEL, shingle, wooden tile
**scirpus, cirpus, sirpus**, rush, rushes
SCOFNETT *see* SCHOFNET
**scolanda, solanda**, prebendal land (of St. Paul's, London)
**scolda**, nagging woman, scold
**scopa 1**, birch tree; besom
**scopa 2**, *see* **shopa**
**scopo**, to sweep
**scoppa**, SCOPE, SCOPPET, SKOPPETT, scoop
**scotala**, scot ale
SCOTE *see* SCHOOT
**scoto, scotio, scottio, shotio, sotio**, to pay scot
**scotum, scotus, scottum**, scot, tax; **s. de capite**, poll-tax; *cf.* **lottum**
**scropha, scroffa**, sow
**scropho**, to root up
**scrophus**, boar

**scrudlanda**, 'shroudland' (providing clothing for monastery)
**sculpo**, to cut; to trim; *cf.* **scapulo**
**scurallum** *see* **curallum**
**scutagium**, scutage, commutation of knight service
**scutella**, dish; bowl
**scutum**, shield
**scyphus, ciphus, cyphus**, bowl; cup
SEADLIP *see* **salopus**
SEAM, SEME, dry measure
**sebum, cebum, cepum**, tallow
**secatura**, cutting; mowing
**seco,** to cut; to mow; to reap
**secta**, suit of court
**sectator**, suitor
**secto**, to sue, to prosecute
**securis**, axe
**secus** *see* **caecus**
SEDKAKE, seedcake (for cattle)
**sedlopum** *see* **salopus**
**seillio** *see* **selio**
**seisina, saisina**, seisin, possession
**seisio**, to take seisin of
**seiso, saisona, seisona, seyso, seysona**, season
**seisono**, to sow (land)
**selda, celda, ceuda, seuda**, stall, shop
**seldagium**, tax on stall
SELFE *see* SHELFE
**seligo** *see* **siligo**
**selio, seillio, sulio, sullio**, selion, strip of land
**sella, cella, scella, zella**, saddle
SELTE, (?) salting
**seltis** *see* **celtis**
SEME *see* SEAM
**semen**, seed
**semilio, semilo**, seedlip
**seminarium, seminatorium**, seedlip
**seminatio**, sowing; sowing season
**semita**, path
**semivirgarius, semivirgatarius**, villein holding half a virgate
**senescalcia**, stewardship
**senescallus**, steward
**separale**, severalty, private holding
**separatio, ceparatio, seperatio**, weaning (time)

**separo**, to separate; to wean
**sepes, cepes,** fence; **s. mortua, 'dead'** (as opposed to 'quick') hedge
**seppagium** *see* **cippagium**
**septum, sceptum**, hedge; enclosure
**sequela**, progeny, young
**sera, cera, serura**, lock; bolt; **s. pendens, pendibilis**, padlock
**seracra, saracra**, SERACRE, acre held by service of carting dung; *cf.* SERLOND
**serculo** *see* **sarculo**
**sericum**, silk
SERLOND, land held by carting service; *cf.* SERACRE
**serra, sarra**, saw
**serro**, to saw
**serura** *see* **sera**
**servagium**, villeinage
**servicium**, tenant service
**serviens**, servant; serjeant
**servilis**, bond, unfree; *see also* **terra**
SESTERNE, cistern
**setto**, to set, to lay
**settatio**, setting (stones)
**seuda** *see* **selda**
**severa** *see* **civera**
**severunda**, CHEVERUND, eaves; cornice
**sextarium**, sester, dry or liquid measure
**seyso, seysona** *see* **seiso**
SHAFTEMUNDE; shaftment, handsbreadth
SHAKETRACE, SHAKETRAYS, SHAKYLTRAYS, (?) chain traces
**sharclutum**, share-clout
**sharrum, scharum**, ploughshare
SHAVE, SAVE, scraper
SHELFE, SELFE, SHELP, shelf; sandbank
SHELVYNGSTOL, cucking-stool
SHEPCOTE, sheepcote
SHEPEN, SHEPENHOUSE, SHEPON, shippon, byre
SHEPHERE *see* SHIPHERE
SHEPSILVER, SCHEPSELVER, money paid in lieu of carrying by water
SHERIFSCHELVER, sheriff's aid, county tax
SHERMAN, shearman
SHETTELLE, shuttle, sluice-gate
SHEWARDE, sea-ward, coastguard duty
SHIFT, to divide
SHIPHERE, SHEPHERE, shiphire, payment for hiring ship

SHOAT, SHOTE, young pig
**shockatio**, stooking (corn)
**shocko, shocco, socco**, to stook
**shockum, shockus, soccus, soka**,
SHOKK, SCHOKK, stook
**shopa, scopa**, shop, stall; *cf.* SOPSELVER
SHORONBRED, (?) protective boarding for mill wheel; *cf.* **shrudo**
SHOTE *see* SHOAT
**shotio** see **scoto**
SHOTSELVER, scot, tax
SHOTTABLE, (?) folding table
**shreddo, †shrido**, to prune, to trim
SHREDYNG, pruning
**shrudo**, (?) to shroud (mill wheel); *cf.* SHORONBRED; SROUDNAYL
**sica**, syke; stream
**siccitas**, drought
**sicco**, to dry
**siccus**, dry; *see also* **precarius**
**sicera, cicera, cisara, cysera, scicera**, cider
SICHEL, sickle
**sigillo**, to seal, to stamp; *see also* **vasum**
**sigillum**, seal, stamp
SIGNET, *see* **cycnettus**
**signo**, to mark; to brand; to stamp
SILEAR *see* SULGHYER
**siligo, seligo, silligo**, rye
**sillum, †zella**, sill; horizontal beam
**silva**, wood; **s. caedua**, coppice wood
**silvestris**, of the woods; wild; *see also* **cattus**
**simila, simula**, fine flour
**simplex**, single; simple; *see also* **hercia**
**simula** *see* **simila**
**sinapium, cinapium**, mustard
**sindula** *see* **scindula**
SINGILNAIL, nail for fixing shingles
**sippus** *see* **cippus**
**sirclum** *see* **circulus**
**sirpus** *see* **scirpus**
**situla**, bucket
SKELLET, SKELLETTE, skillet
SKOMERE, SKUMURE, skimmer
SKOPPETT *see* **scoppa**
SKUMURE *see* SKOMERE
SLADE, SLEDE, valley; hollow; glade
SLATTERSHEEP, sheep set aside for slaughter

SLEDE *see* SLADE
SLIPE, SLYPE, narrow piece of ground
SLO, SLOW, slough, mire
SLOTERE, (wooden) cross-piece
SLYPE *see* SLIPE
SMOKE HEN, customary due; (?) reek hen
SNACCHE, SNEKE, hasp; snare (to catch dogs in warren)
SNAKEWERE, kind of weir; *cf.* **wera**
SNARTH, snath, shaft of scythe
SNEKE *see* SNACCHE
SNOTTERINGSILVER, customary rent (paid to Colchester Abbey); *cf.* KNOWLEDGE MONEY; **recognitio**
**soca, soka**, soke; suit of court; *cf.* **sokena**
**socagium**, socage, tenure involving suit of court
**socco** *see* **shocko**
**soccus 1** *see* **shockum**
**soccus 2**, ploughshare
**soka 1** *see* **shockum**
**soka 2** *see* **soca**
**sokerellus**, (?) suckerel, suckling
**sokemannus**, sokeman; tenant in socage
**sokena**, soke; *cf.* **soca**
SOKNEBEDRIPE, harvest boonwork
**solanda** *see* **scolanda**
**solarium**, solar, upper room
**soldatio, soldatura**, soldering
**soldo, soudo, zoudo**, to solder
SOLE, SOULE, SOULLE, sill
**soliata** *see* **terra**
**solidata 1**, shillingsworth, land yielding one shilling rent
**solidata 2** *see* **terra**
**solidus 1**, solid; undisturbed
**solidus 2**, shilling
**soliva** *see* **sulliva**
**solta**, payment
**solum**, ground; base
**solus**, alone
SOMER, (?) supporting beam
SOMERHEGGE, summer hedge
SOMPE, SOMPTE, sump; swamp; marsh
SOMURREP, summer harvest
SOPSELVER, SOPESELVER, (?) rent for shop or stall; *cf.* **shopa**
SORELLBALD, skewbald

SORFTRE *see* CERFTRE
**sotio** *see* **scoto**
**sotularis**, shoe, slipper
**soudo** *see* **soldo**
SOULE, SOULLE *see* SOLE
SPAN, SPANG, SPONGE, fastening; link
SPANSHAKEL, clasp on plough chain
**spargettor**, sower
**spargo**, to scatter, to spread; to ted (hay)
**sparra, sparrum**, rafter; beam
**sparsio**, scattering, spreading
**sparsorius**, (?) seedlip
**spatior**, to wander
**spatium**, space; (?) work break
SPELDE, splinter, chip
SPENCE, SPENSE, pantry, larder
SPERE, SPIRE, sapling
SPEREBOURNEWODE, sapling wood
SPERGINPEK, SPRENGYNPEK, seedlip; *cf.* SPRENGERE
SPERTE *see* **sporta**
**spervarius**, sparrow-hawk
**spica**, ear of corn
SPIKE, SPIKING, SPIKYNG, SPIKYNGEL, SPYKYNG, spike-nail
**spina**, thorn; thorntree
SPIRE *see* SPERE
**spissitudo**, thickness
**spissus**, thick
**spiteclutum**, SPITCLUT, SPYTCLOUT, iron reinforcement on plough
**splenta**, SPLENTE, 'splint', lath
**splentatio**, fitting of laths
**splento, splinto, splyynto**, to fit w. laths
**spoka**, spoke
SPONGE *see* SPAN
SPORE, spur
**sporta**, SPERTE, basket
**sprendla, sprengula**, SPRENDLE, SPRYNDELE, split rod; thatching-peg
SPRENGERE, seedlip
SPRENGYNPEK *see* SPERGINPEK
**sprengula** *see* **sprendla**
SPRIG, SPRYG, (?) mixed or early malting grain
SPRIGNAIL, small nail, brad
SPRING, copse of young trees
SPRYG *see* SPRIG
SPRYNDELE *see* **sprendla**

SPYKING *see* SPIKE
SPYTCLOUT *see* **spiteclutum**
**squarratio**, squaring
**squarro**, to square (timber)
SROUDNAYL, shroud-nail; *cf.* **shrudo**
**stabularius**, stableman
**stabulum**, stable
**staca**, stake
**stacco, stakko**, to stack
**staccus**, stack
**staddellus, stathelus**, STALLYNG, STATHELE, staddle, tree left in cleared woodland
**stadium**, measure of length, esp. furlong
**stafizacra**, †STAFACRE, STAVESACRE, plant used as emetic
**stagneus**, made of tin or pewter
**stagnum 1, stannum**, tin, pewter; *cf.* **stannatus**
**stagnum 2, stangum**, pond; millpool
**stakko** *see* **stacco**
**stallagium**, stallage, fee for market stall
**stallum, stallus**, market stall
STALLYNG *see* **staddellus**
**stalo**, stallion
STALPLACE, place for a stall
STAMPYNGTROW, fuller's or dyer's trough
**standardum**, upright timber, post
**stangum** *see* **stagnum 2**
**stannatus**, covered w. tin; *cf.* **stagnum 1**
**stannum** *see* **stagnum 1**
†**stantivo**, (?) to prop up
**stantivus**, STANTIF, upright
STAPELE, (wooden) bar, staple
**stathelus**, STATHELE, *see* **staddellus**
**statio**, standing idle (of mill)
**staurum**, store, stock; **s. mortuum**, non-live stock, **s. vivum**, livestock
STAVESACRE *see* **stafizacra**
STELE, long handle
STELEWERE, kind of weir; *cf.* **wera**
STELHOUS, stable, outhouse
**stercorarium**, privy; dunghill
**stercoratus**, manured
**sterculenum, sterculinum, sterculunum, sterquilinum**, dunghill
STERKE, stirk, heifer
**sterquilinum** *see* **sterculenum**
STERTE, start, inner section of mill-wheel bucket

**stica, estaka**, 'stick', measure of eels
**stillicidium**, gutter; eaves
**stimulus**, goad
**stipes**, trunk; post; *see also* **par**
**stipula, stupula**, stubble, straw
**stipulo**, to gather stubble; to thatch
STIROP, stirrup; S. **ferri**, attachment to cogwheel of mill
**sto**, to stand; to stay
**stocco** *see* **stocko**
**stockingum**, clearing (of woodland)
**stocko, stocco**, to stub up (woodland)
**stoda, stodum**, *see* **studa**
STORVEN, dead; (?) lopped
STOTHE *see* **studa**
**stothio** *see* **studo**
STOTSCHEPENE, byre for stots
**stottus**, stot, steer, horse; plough-beast
STRADEL *see* **stradulum**
STRADELCLUT, STRADECLUT, STRADILCLUT, iron plate protecting 'stradel'
STRADELENCS, links for 'stradel'
**stradulum**, STRADEL, (?) end of axle
**stragulatus**, striped
**straia** *see* **estraia**
STRAKBANDE, iron strap
STRAKE, STRAKKE, 'streak', iron strip on wheel
STRAKNAIL, nail for fixing 'streak'
**stramen**, straw
STRAUNGELOUN, strangullion, strangles, disease of horses
STREDDORE, street (front) door
STREKE 1, STRIKE, to 'strike', to furrow
STREKE 2, *see* **strica**
†**strekeo**, (?) to stretch (straw for thatch)
STRENOURCLOTH, straining-cloth
STREYDERE, sower
STREDYNGPEK, seedlip
**strica, stricus, strika**, STREKE, 'strike', dry measure; bundle of hemp; *cf.* ESTRICKE
**strigilis**, comb, currycomb
**strika** *see* **strica**
STRIKE *see* STREKE 1
**strippum** *see* **estrepum**
**stroda**, STRODE, 'strood', marshy land
**stubbatio, stubbatura**, stubbing

**stubbo**, to stub-up
**stubbum**, stub; stump
**studa, stoda, stodum**, STOTHE, stud, upright in wattle wall
**studo, stothio**, to fit walls w. studs
STUE, ETUE, stew, fishpond
**stuffatus**, stocked
**stuffum, stuffura, estuffum**, store, stock
**stulpa, stulpis**, post; whipping-post
**stuppo, estoppo**, to stop-up; *cf.* **estupamentum**
**stupula** *see* **stipula**
**suarium, suaria**, pig-stye; pig pasture
**subancilla**, under-maidservant
**subballivus**, under-bailiff
**subboscus**, underwood, brushwood
**subconstabularius, subconestabulus**, under-constable
**subfugator**, under-drover
**subpedito** *see* **suppedito**
**succidium**, souse, pickled (pork)
**succido**, to cut down; (?) to cull
**succisio**, cutting down
**succus**, juice
**suffocalia**, (?) bellows
**sulco**, to furrow
**sulcus**, furrow; **s. aquaticus**, drain
SULGHYER, SILEAR, SULEER, SULGTRYER, SWILLYARD, part of plough, (?) notched vertical iron at front
**sulio, sullio**, *see* **selio**
**sulliva, soliva, sullivum**, sill
SULSHO, (?) ploughshoe
**summa 1**, sum, total
**summa 2**, load
**summagium**, carrying service
**summarius**, pack-horse
**summoneo**, to summon
**summonitio**, summons
**superlectum, superpelectulum** *see* **suppellectile**
**superoneratio**, overstocking; overcharging
**superonero**, to overstock; to overcharge
**supersolum**, bank, dam
**supertunica**, surcoat
**suppedito, subpedito**, to trample; (?) to reinforce, to underpin

**suppellectile, superlectum, superpelectulum**, bedding; coverlet
**suppodio**, to prop
**suppodium**, (?) support
**sursisa**, SURSYSE, (fine for) default
**sus**, sow
**susenna**, upland
**suspendo**, to hang
SWATHE, SWAD, swathe, space covered by sweep of scythe
SWENGYNGHOK, (?) scythe
SWEPE, sail (of windmill)
SWEVELE, swivel
SWILLYARD see SULGHYER
SWYCH, switch, slender rod
SWYNHOUS, pig-stye
SYKEL, sickle; (?) brace on sail of windmill
SYNGUL see **cingulum**

# T

TAAR, THAR, tar
**tabella**, board; sideboard; **tabellae**, 'tables', (?) backgammon
**taberna**, tavern
**tabernarius, tabernator**, taverner; (?) tavern-haunter
**tabula**, board, table; **t. clausa**, 'close', boarded table; **t. dormiens**, fixed table; **t. mensalis**, trestle table, **t. pedalis**, (?) footboard
**tabulatus**, boarding
**tabulo**, to board up
**tachio, thachio**, to tether; to attach
**taillia** see **tallia**
TAINTERFIELD, TENTORCLOSE, tenterfield, place for drying cloth
TALEWODE, TALLOWODE, wood cut to size
**tali** (*pl.*) dice
**tallagium**, tallage, tax
**tallia, taillia**, tally
**tallio**, to tallage
TALLOWODE see TALEWODE
**talpa**, mole; *see also* **collis; mons**
**tannator**, tanner
TAPPEHOSE, strainer
TAPPETROWE, trough used in brewing
**tarambium**, gimlet, auger
**tarpica**, tar-pitch; (?) turpentine
**tasca, tascha**, piece-work
**tascator, taxator**, piece-worker; tax-assessor
**tassa, tassum**, stack; haycock
**tassatio**, stacking
**tassator, tassor**, stacker
**tasso**, to stack
**tastator**, taster; **t. cervisiae**, ale-taster
**tasto**, to taste; to inspect
TATTOCK, (?) fishhook
**taurellus, tauriculus, tauriolus**, steer, young bull
**taurus**, bull
TAWYING, dressing leather
**taxator** see **tascator**
TAYLROPE, breech-rope, part of harness
**tector**, roofer; thatcher
**tectura**, roof; roofing
**teddo**, to spread hay
TEDDYING, haymaking, hay spreading
**tegula**, tile
**tegulator**, tiler
**tegulo**, to tile
**teisa, teysa**, fathom (measure)
**tela 1**, TELE, teal
**tela 2**, cloth; sheet
**telarius**, weaver
TELE see **tela 1**
**teloneum** see **theoloneum**
**temo, themo**, cart-shaft; plough beam
**temptatio**, testing
**temptator** see **tentator**
**tempto** see **tento**
**tempus**, time; season; **t. apertum**, open season for grazing; **t. pestilenciae**, the Black Death; **t. rumoris**, the Peasants' Revolt
**tenator** see **tentor**
**tencha, tenchea**, tench
**tenementum**, tenement, holding

**teneo**, to hold; *see also* **copia**
**tentator, temptator**, tester; **t. cervisiae**, aletaster
**tentio**, holding
**tento, tempto**, to try; to test
**tentor, tenator**, ploughman
TENTORCLOSE *see* TAINTERFIELD
**tenura**, holding
**tera**, tether
**terra**, land, esp. arable; **t. frisca**, fresh (not salt) land; **t. nativa, servilis, villana**, bond land; **t. solidata, soliata**, 'soiled' land, freehold bought by bondman
**terrarius**, tenant
**terrenus** (*adj.*), land-holding
**terrula**, small piece of land
**textor, textrix**, weaver
TEYBACE, (?) support for tie-beam
TEYLNAYL, tile-nail
**teysa** *see* **teisa**
**thacio** *see* **tachio**
THAR *see* TAAR
**theca**, chest; (money) box
THECCHERE, thatcher
THELLE, (?) board, plank
**themo** *see* **temo**
**theoloneum, teloneum**, toll
THEVE, theave, young ewe
THEWE, pillory for females
THILL, THYLLE, cart-shaft
THILLEHAME, loop supporting cart-shaft
**thixillum**, THIXEL, adze
**thorale** *see* **torale**
THORNSTYKKE, blackthorn, cudgel
**thrava, trava**, shock (of corn)
THURSCHOTE, opening (for sail of mill)
THYLLE *see* THILL
TIGILPYNNE, TILPYNNE, tile-pin
**tignum**, beam
TILPYNNE *see* TIGILPYNNE
**tina**, tub; cask
**tinctor, tinxtor**, dyer
**tinctorium**, dye-works
**tinda, tynda**, tine, tooth
**tindo, tineo, tino**, to fit w. teeth
**tinxtor** *see* **tinctor**
**tipulator, typpelator**, 'tippler', retailer of ale
**tipulo**, to sell ale
**tirieca, triaca**, theriac, 'treacle'
TOFSCHEF, sheaf given to toft-holder
**toftum**, toft; house-site
**toga**, gown
TOLHUCCHE, toll-box
**tollagium**, toll
TOLLTRAY, TOTTERAY, toll on corn
**tolnetum**, toll
**tolsester**, toll of one sester (of ale); *cf.* CUSTOMPOTTES
**tomerellum** *see* **tumberellum**
**tondeo**, to shear
**tonella, tunella**, small cask
**tonna**, tun, cask
**tonsio, tonsura**, shearing
**torale, thorale**, kiln
TORCHWEKE, torch-wick
**torculare**, (cider) press; *see also* **domus**
**tornatilis** *see* **bancus 2**
TOTENEYS, cloth from Totnes (Devon)
TOTTERAY *see* TOLLTRAY
TOWEYREN, (?) tew-iron, part of bellows or smelting furnace
TOYLE, toil, snare
**trabs, trava**, TRAVE, beam, shaft
**tractator**, (ale) drawer, tapster
**tractus**, sweep (of scythe); trace (of harness)
**traho**, to draw (out); **t. moram**, to abide
TRAMALE, TRAMAYLLE, trammel, fishing-net
TRAMAYS *see* **tremagium**
TRANSTAVE *see* **trendstavum**
**trava 1** *see* **thrava**
**trava 2**, TRAVE *see* **trabs**
TRAVIS, open shed (of smithy)
**trebuchettum, trebechetum, tribuchetum**, cucking-stool
**tremagium, tremasium, tremesium**, TRAMAYS, summer-sowing
TRENCHES, trench, colic (of horse)
**trendella**, TRENDLE, TRENDYLE, trendle-wheel (of mill)
**trendstavum**, TRANSTAVE, TRENDLESTAVE, stave on trendle
**treparium** *see* **tripes**
**tresantia, trisantia**, TRESAUNCE, corridor, screened passage

TRESSHOLT, threshold
**trestellum, trestallus**, TRISTALL, trestle
**triaca** *see* **tirieca**
**tribuchetum** *see* **trebuchetum**
**tribula, tribulus**, shovel; fork
**tribulus**, thorn, briar
**tripes, treparium**, tripod; trivet
**tripudio, tripido**, to dance on; to flatten
**trisantia** *see* **tresantia**
TRISTALL *see* **trestellum**
**trituratio**, threshing
**trituro**, to thresh
TROWE, trough
**truia**, sow
**truncus**, tree-trunk; block of wood
**trussellus**, bundle
TRUSSING BED, folding bed
**trusso**, to pack; to dress
**tumberellum, tomerellum**, tumbrel, farm cart

**tunella** *see* **tonella**
**turba**, turf; peat
**turbarium**, turbary
**turnarium, turnera**, headland, turning-place
**turnus (vicecomitis)**, sheriff's tourn
**turpo**, to befoul
TURTEL, TURTHEL, (?) tortoiseshell (coloured)
**turvera**, turbary
TUSSERDE, TUSARDE, faggot
TWELE, twill
TYE, strip of (enclosed) land
TYLEKYLL, tile-kiln
**tynda** *see* **tinda**
TYNGELNAIL, small nail, tack
**typpelator** *see* **tipulator**

# U

**ulmus**, elm tree
**ulna**, ell
UMPLE, fine linen cloth
**unae** *see* **unus**
**unctum, unctura**, oil; grease; ointment; *see also* **panis**
UNDERCLUT, underclout (of plough)
UNDERLAWE, underlay; (?) foundation
UNDONKESAKER, UNTHANKESAKER, harvest custom
UNGELD *see* UNYELDE
**unus**, one, an; **uni, unae**, a pair
UNYELDE, HUNTHIELD, ONGIELL, UNGELD,

UNTHIELD, UNZELD, rent (?) in lieu of hunting service
**urbanitas**, kindness; service;
  **u. Angliae**, courtesy of England (form of tenure)
**urceolus, urciolus**, pot, pitcher
**ustilamenta, ostilmentum**,
  HOSTYLEMENTS, household goods
**ustrinum**, (malt) kiln
**utlagatus, hutlagatus, utlegatus**, outlaw
UTWODE, outlying woodland; *see also* **boscus**
**uva**, grape

# V

**vacans**, empty
**vacca**, cow
**vaccaria**, cow-pasture; cow-shed
**vaccarius**, cowman
**vadelectus** *see* **valettus**
**vadio**, to pledge
**vadium**, bond, pledge

**vaga, vagum**, waif, chattel abandoned by felon
**valettus, vadelectus, valectus**, groom
VALWE, fallow, brownish-yellow
**vanga**, spade; shovel
**vanna, fanna, vannus**, winnowing fan
**vannatio**, winnowing

45

VANNE see **fennum**
**vanno, fanno**, to winnow
VARN, fern, bracken
**vastum, wastum**, waste (land); damage
**vasum**, bowl, vessel; **v. apium**, beehive, **v. non sigillatum**, illegal (unstamped) vessel
**vatta, factus, fattum, vattum**, FATTE, vat, tub; (?) measure
**vectura**, transport; baggage
VEDFEE, merchet, payment by villein on daughter's marriage
VEERNE see **verna**
**velamen**, sail; veil; sheet
**velia** see **felea**
VELIE, VELY see **felea**
**vellus**, fleece
**velum**, sail; veil; sheet
**venella**, lane
**venter**, belly; see also **lana**
**ventilarium, ventilabrum**, winnowing fan
**ventilo, ventulo**, to winnow
**ventricius**, driven by wind; see also **molendinum**
**ventus**, wind; see also **molendinum**
**vepris**, thorn bush
**verbero**, to beat; to thresh
VERDEGRIZ, verdigris
**verdera** see **viridarium**
VERDLINGGE, VERDLYNGE see **ferlingus**
**vermina**, vermin
**vermis**, worm
**verna**, VEERNE, windlass
**verres, verrus**, boar
**verro**, to grub-up
**vertivellum, vertevellum**, hinge
**verto**, to turn; to make (hay)
VERULE see **virolla**
**vervex**, wether
**vesca**, vetch
**vestitus, vestura**, clothing; sail (of windmill); crop (of corn)
VETTERLOK see FETLOCK
**via**, road, way; **v. carrettae**, cartway; **v. communis**, public road; **v. equestris**, bridlepath; **v. fugalis**, droveway; **v. pedalis, pedestris**, footpath; **v. regia**, highway

**vicecomes**, sheriff; see also **turnus**
**victus**, food, sustenance
**vicus**, street
**vigil**, watch; guard
**vigilo**, to keep watch
**villa, villata**, 'vill', village, township
**villanus**, villein, customary tenant; see also **terra**
**vinea**, vine; vineyard
**viniator, vinitor**, vintner; vinedresser
VINSPARRES see FYNSPARRIS
**vinum**, wine
**virga**, rod; yard (3 feet); sailyard; withy
**virgarius, virgatarius, virgator**, tenant of virgate
**virgata**, virgate, quarter-hide; cf. **hida**
**virgulta, virgultum**, brushwood
**viridarium, verdera**, green place
**viridarius**, verderer, forest official
**virolla**, VERULE, ferrule; boss
**visnetum, visinetum**, neighbourhood
**vitelarius** see **vitularius**
**vitrarius**, glazier
**vitro**, to glaze
**vitularius, vitelarius**, victualler
**vitulabilis**, (?) able to calve
**vitulatio**, calving
**vitulo**, to calve
**vitulus**, calf; see also **jacto**
**vivarium**, fishpond, stew
**vivus**, alive; see also **argentum; pecunia; staurum**
**vixo** see **wisco**
**voga, wogium**, pruning-knife; bill-hook
**volatile**, fowl; **v. warennae**, game bird
**vomer**, ploughshare
VTERLOK see FETLOCK
**vulpes, wlpes**, fox
VYNDYNGBEND, winding-band; (?) tyre of wheel
VYNSPARRIS see FYNSPARRIS
VYNYERD, VYNEYERD, vineyard
VYTHELBORD see **fetherborda**

# W

WACHET, watchet, light blue cloth
WADMOL, wadmal, coarse woollen cloth
**waga, waya, weia**, WAYE, WAYGH, WEY, 'wey', measure of weight (esp. of cheese)**wagessa**, WAGESSE, swamp; (?) water-weed
**wahura** *see* **wayarium**
**waida, weida**, WEWDE, woad
**wainabilis, wainnabilis**, cultivable, profitable
**wainagium, gainagium, gaynagium**, cultivation; means of cultivation; farm stock
**waineria, gayneria**, yield (of cultivation)
**wainum**, WEYN, wain, wagon
**waivio, weivio, weyvo**, to abandon; to forfeit (chattel)
**waivium**, WAYF, WEYF, abandoned chattel
**walda** *see* **walla**
**waldura**, walling; building of embankment
WALGAVEL, (?) tax for maintaining town wall
**walla, walda, wallea**, wall, esp. sea-wall; embankment
**Wallia**, Wales; *see also* **nux**
**wallura**, wattling; *cf.* **watillo**
WAMBELOKES, WAUMBELOKKES, sheep's belly-wool; *cf.* **lana**; BELLEFLES; WEMBS
WANT *see* WENT
WANTEY, WANTEYE, wanty, belly-band
**wara 1**, unit of geld
**wara 2** *see* **wera**
**warda**, guard, ward
**wardacra**, land reaped in lieu of ward service; *cf.* **acrewara**
**wardarius, garderius**, keeper, guardian
WARDEPENI, WERPENI, ward-penny, payment in lieu of ward service; *see also* KYNGGESWARD
WARDESCOT, WARDESHOT, ward-penny
WARDESELVER, ward-penny
WARDSTAFF, WARDESTOF, willow wand symbolising the King's Peace

**warecta, warectum**, fallow (land); *see also* **jaceo**
**warecto**, to fallow
**warenna, garenna, garuna**, (rabbit) warren; **libera w.** right of free warren; *see also* **volatile**
**warennarius**, warenner
WAREPAILL, bucket
**warra** *see* **wharva**
**warrecum** *see* **wreccum**
**warva** *see* **wharva**
**wastellus, gastellum, wastellum**, wastel-bread
**wastum** *see* **vastum**
WATERBEDRIP, boon-reaping without ale
WATERDORE, sluice-hatch
WATERENFURW, WATERFORE, WATERFOROUGH, drainage furrow
**waterfurgio**, to make drainage furrows
WATERLANE, green lane with stream; (?) millrace
WATERLOCK, lock; dam
WATERSCHIPPE, (?) ferry boat
WATERSELLE, (?) pond
WATERWOU, WATERWOUGHE, WATERWOW, waterwall, sides of millrace
**watillo**, to wattle; *cf.* **wallura**
WATIRFOURSHOVELE, spade for digging drainage furrows
WAUMBELOKKES *see* WAMBELOKES
WAXSILVER, WEXSILVYR, levy for church candles
**waya** *see* **waga**
**wayarium, wahura**, WAYERE, WAYOUR, horse-pond
WAYE *see* **waga**
WAYERE *see* **wayarium**
WAYF *see* **waivium**
WAYGH *see* **waga**
WAYLET, WEYLATE, cross-roads
WAYNSKOTT, high quality oak board
WAYOUR *see* **wayarium**
WDERICHT *see* WODERADE
WDESELVER *see* WODESELVER
WEDERBEME, weather-beam (of windmill)
WEDERE, one who weeds

# WEDHOKE

WEDHOKE, WYDHOKE, weeding-hook
**weia** *see* **waga**
**weida** *see* **waida**
**weivio** *see* **waivio**
WEKE, (candle) wick
WELE, WHELE, trap, esp. for eels; basket
WEMBS, (?) belly-leathers; *cf.* WAMBELOKES
WENNEL, weanel, young calf
WENT, WANT, way, road; *see also* FOURWENT
WENYARD, whinyard, short sword
WEPILTREVIS, WEPYTRE, whipple tree (on plough)
**wera, wara**, WERE, weir; *see also* CROUCHWERE; EBWERE; FLODWERE; SNAKEWERE; STELEWERE
**werda**, WERDE, WEYERD, marsh
**werkfanno, workfanno**, to winnow
WERKHOUS, workshop
WERPENI *see* WARDEPENI
WETHERHERDE, shepherd in charge of wethers
WEWDE *see* **waida**
WEXSYLVYR *see* WAXSILVER
WEY *see* **waga**
WEYERD *see* **werda**
WEYF *see* **waivium**
WEYLATE *see* WAYLET
WEYLESE, (?) way-leave
WEYN *see* **wainum**
WEYSHYDES, (?) planks lining mill-sluice
**weyvo** *see* **waivio**
**wharva, quarva, warra, warva**, wharf; quay
**wharvagium**, wharfage
WHARVING, wharfing; (?) scaffolding
WHELE *see* WELE
WHETEBERNE, wheat barn
**whola, wolvus**, WHOLME, WHOLVE, WHULF, WOLF, WOLVE, wholve, arched drain (often under gateway); *see also* OLVE

**wica**, WICK, WYK, wick, dairy-farm; (?) sheep walk
WIGG, 'wig', bun, small cake
**windagium, wyndagium**, fee for hoisting
**windasius, wyndasius**, WYNDAS, windlass
WINDBRECHE, windfall wood
**wippa**, whip; wisp; bundle
WIRPES, (?) murrain (of deer)
**wisca**, wattle
**wisco, vixo, wixo**, to wattle
**wlpes** *see* **vulpes**
WODECOKK, WODECOTT, woodcock
WODERADE, WDERICHT, WODERIDTE, right to take wood
WODESELVER, WDESELVER, payment in lieu of carrying wood
WODEWARDE, woodward
**wogium** *see* **voga**
**wolvus**, WOLF, WOLVE *see* **whola**
WOMBEROPE, WOMBETYE, WOMECLOUT, girth, belly-band
**workfanno** *see* **werkfanno**
**wosa**, WOSE, mud; ooze
**wreccum, warrecum, wrekkum**, wreck; wreckage
†**wrevia**, (?) iron hoop on hurdle
**wroto**, to root (of pigs)
WYDHOKE *see* WEDHOKE
WYK *see* **wica**
WYLECALTRE, (?) (eel) trap
**wyndagium** *see* **windagium**
**wyndasium**, WYNDAS *see* **windasius**
WYNDBEME, wind-beam, cross-beam tying roof-rafters
WYNDYNGRODDE, thatcher's rod
WYNHUSE, wine-store; (?) public house
WYNTERSTORE, keep, winter-fodder

# Y

**yda** *see* **hida**
YEANING, lambing; *cf.* YENYNGCORN
YELMING *see* HAULMING
YELT *see* GILT
**yemalis** *see* **hiemalis**
YENYNGCORN, ENYNCORN, extra feed for lambing ewes
YERDLING, ERDLYNG, (?) yardland; *cf.* **ferlingus**
YEUMAN, yeoman, hind
YEVEHANN, (?) service of hen given at Christmas
YEVESHEP, (?) service of hogget given at Whitsun
YEVEWERKE, boonwork
YMPE, 'imp', sapling; *cf.* **impo**

**yncherna**, YNCHERNE, (?) inner structure of dovehouse
YNHUWE *see* INHEWE
YNNED *see* INNED
**yvernagium** *see* **hibernagium**

# Z

**zabulum** *see* **sabulum**
†**zella** *see* **sella**; **sillum**
**zinziber, gingiber**, ginger; *see also* **racemus**
**zona**, belt, girdle

**zoudo** *see* **soldo**